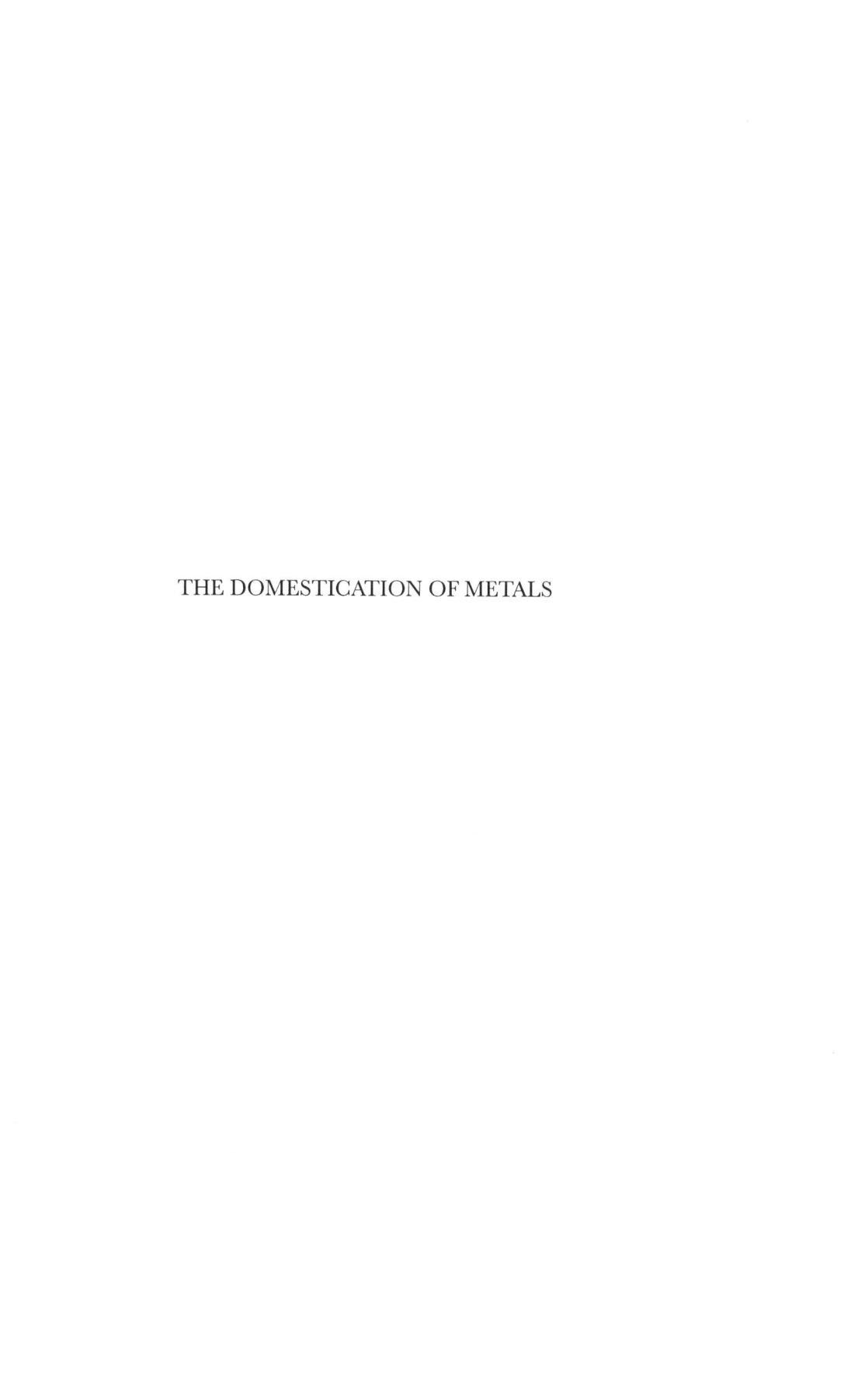

THE DOMESTICATION OF METALS

CULTURE AND HISTORY OF THE ANCIENT NEAR EAST

EDITED BY

B. HALPERN, M. H. E. WEIPPERT
TH. P.J. VAN DEN HOUT, I. WINTER

VOLUME 4

THE DOMESTICATION OF METALS

The Rise of Complex Metal Industries in Anatolia

BY

K. ASLIHAN YENER

BRILL
LEIDEN · BOSTON · KÖLN
2000

This book is printed on acid-free paper.

Library of Congress Cataloging-in-Publication Data

Yener, K. Aslihan.
　The domestication of metals : the rise of complex metal industries in Anatolia / by K. Aslihan Yener.
　　p. cm.—(Culture and history of the ancient Near East; ISSN 1566-2055 ; v. 4)
　Includes bibliographical references and index.
　ISBN 9004118640 (hc : alk. paper)
　1.Copper age—Turkey. 2. Bronze age—Turkey. 3. Metal-work, Prehistoric—Turkey. 4. Turkey—Antiquities. I. Title. II. Culture and history of the ancient Near East.
GN778.32.T9 Y45　　　2000
939'.2—dc21
　　　　　　　　　　　　　　　　　　　　　　　　　　　　00-026377
　　　　　　　　　　　　　　　　　　　　　　　　　　　　CIP

Die Deutsche Bibliothek – CIP-Einheitsaufnahme

Yener, K. Aslihan :
The domestication of metals : the rise of complex metal industries in Anatolia / by K. Aslihan Yener. – Leiden ; Boston ; Köln : Brill, 2000
　(Culture and history of the ancient Near East ; Vol. 4)
　ISBN 90-04-11864-0

ISSN　1566-2055
ISBN　90 04 11864 0

© Copyright 2000 by Koninklijke Brill NV, Leiden, The Netherlands

All rights reserved. No part of this publication may be reproduced, translated, stored in a retrieval system, or transmitted in any form or by any means, electronic, mechanical, photocopying, recording or otherwise, without prior written permission from the publisher.

*Authorization to photocopy items for internal or personal use is granted by Koninklijke Brill provided that the appropriate fees are paid directly to The Copyright Clearance Center, 222 Rosewood Drive, Suite 910, Danvers MA 01923, USA.
Fees are subject to change.*

PRINTED IN THE NETHERLANDS

TABLE OF CONTENTS

List of Tables vii

List of Figures viii

List of Plates x

Acknowledgements xi

CHAPTER 1. THE RISE OF COMPLEX METAL INDUSTRIES IN
ANATOLIA, ANCIENT TURKEY

    Introduction..1
    The Intellectual Framework 4
    Metal Production in Highland Anatolia:
    Innovation at the Frontier..10
    Case Studies of Production Models 12

CHAPTER 2. ARCHAEOLOGICAL BACKGROUND

    Introduction..17
    The Technology of Prestige:
    The Aceramic and Pottery Neolithic Beginnings........................18
    Transformations in Technology and Organization in the
    Chalcolithic Period (c. 5500-3000 B.C.)....................................25
    A. The Ubaid Period (late 5th and early 4th millennium B.C.) 30
        Case Study 1: Değirmentepe (Malatya)
    B. The Technology of Prestige and Power: The Uruk Contact
    (c. 3400-2900 B.C.)..44
        Case Study 2: Arslantepe
        The Altınova Valley Sites: Keban Dam Salvage
           Projects
        The Mediterranean Coast
    C. The Early Bronze Age: Industrial Production.........................67

CHAPTER 3. KESTEL MINE AND GÖLTEPE

    The Problem of Tin Sources...71

Field Research in the Central Taurus Mountains:
The Physical Setting 76
The Bolkardağ Area 76
    The Bokardağ Area Site Survey
The Çamardı Area 80
    The Çamardı Area Site Survey
The Kestel Intensive Surface Survey 85
    Sounding S.B.
Excavations at Kestel Tin Mine 88
    Kestel Mine Soundings S.1-S.4
    Burial Chambers
Intensive Surface Survey at Göltepe 98
Göltepe, Tin Smelting Workshops, and Habitation 101
    Area A and Area B Pithouse Structures

CHAPTER 4: THE PRODUCTION OF TIN

The Smelting Process 111
Ore Materials from Göltepe 112
Analysis of the Earthenware Crucible/Bowl Furnaces 115
Smelting Experiments 121

CHAPTER 5: CONCLUSIONS 125

BIBLIOGRAPHY 129

INDEX 161

TABLES 169

FIGURES 183

PLATES 213

LIST OF TABLES

Table 1: Trace element analyses of Bolkardağ ores and slag samples.
Table 2: Çamardı sites.
Table 3: Kestel radiocarbon dates.
Table 4: Göltepe radiocarbon dates.
Table 5: Atomic absorption analysis of metal objects from Göltepe.
Table 6: Comparison of the elemental analysis of hematite samples.
Table 7: Atomic absorption analysis of Göltepe powdered samples.
Table 8a: General sample information;
8b: Elemental concentrations measured by X-ray fluorescence.
Table 9: Average composition of the tin-containing particle types in the nine powder samples.
Table 10: XPS results.

LIST OF FIGURES

Fig. 1: Map of Turkey.
Fig. 2a: Histogram of arsenic content in Near-Eastern copper and bronze objects;
    2b: Histogram of arsenic content in tinless copper objects. (Period 2, late 4th-early 3rd; Period 3, late 3rd; Period 4, Middle Bronze Age; from Caneva, Frangipane and Palmieri 1985: 128, Fig. 6)
Fig. 3: Distribution of metallurgical debris from Değirmentepe (after Esin 1989).
Fig. 4a: A bimodal distribution is indicated for the differences between arsenic contents of swords versus spears (from Caneva, Frangipane, and Palmieri 1985: 117);
    4b: A ternary diagram of the trace elements in the artifacts suggests that most were derived to a lesser extent from oxides and sulfides (from Caneva and Palmieri 1983: 643).
Fig. 5: Metal content of ores found at Arslantepe (from Palmieri, Hauptmann, Hess, and Sertok 1996: Fig. 1).
Fig. 6: Topographical map of the Bolkardağ region.
Fig. 7: Topographical map of the Celaller-Kestel region.
Fig. 8: Distribution of sites in the Niğde Massif.
Fig. 9: Kestel Mine slope (Sarıtuzla) workshop and mine entrances (1988 survey).
Fig. 10: Density map of ceramics from Kestel Mine slope survey.
Fig. 11: Large ore-dressing installation at the entrance of Kestel Mine.
Fig. 12: Artifacts from Göltepe and Kestel Mine.
Fig. 13: Artifacts from Kestel Mine.
Fig. 14: Kestel Mine ceramic assemblage.
Fig. 15: Plan of Kestel Mine (Lynn Willies).
Fig. 16: Plan of mortuary chamber, Kestel Mine.
Fig. 17: Groundstone tool distribution map (Göltepe survey).
Fig. 18: Summit map of Göltepe.
Fig. 19: Excavation trench map, Göltepe.
Fig. 20: Ceramic molds for a flat ax and chisel (Göltepe, Early Bronze Age).
Fig. 21: Pithouse structures 6 and 15 (Göltepe, Early Bronze Age).
Fig. 22: Structures B05 and B06 in Area B (Göltepe, Early Bronze Age).
Fig. 23: Silver necklace (Area B, Göltepe, Early Bronze Age).

LIST OF FIGURES

Fig. 24: Crucibles (Göltepe, Early Bronze Age).
Fig. 25: Plan of Area E showing midden deposits.
Fig. 26: Chart of abundance of particle groups in nine powder samples.
Fig. 27: Chart of abundance of tin-containing particle groups in nine powder samples.

LIST OF PLATES

Plate 1: Areal view of Cilicia, Bolkardağ, and Çamardı (M.T.A.; December 16, 1972).
Plate 2: Cassiterite grains panned out of the Kuruçay Stream (Necip Pehlivan, M.T.A.; 1987).
Plate 3: Computer model of Çamardı, Celaller, Kestel Mine and Göltepe region (John and Peggy Sanders, Oriental Institute).
Plate 4: Crucible fragments from Kestel slope area.
Plate 5a: Chamber VI Kestel Mine;
5b: Cassiterite grain from Sounding S2 (Chamber VI, Kestel Mine).
Plate 6a: Dark-burnished ware (Chamber VI, Sounding 2);
6b: Straw-tempered ware (Chamber VI, Sounding 2).
Plate 7: Antler tools from Kestel Mine.
Plate 8: Decorated ceramic panel (Pithouse 6, Göltepe, Early Bronze Age).
Plate 9: Large storage vessel containing ground ore material (Pithouse 6 Göltepe, Early Bronze Age).
Plate 10: Large crucible with stone covers on floor of Pithouse 15 (Göltepe, Early Bronze Age).
Plate 11: Geometrically decorated ceramic panel over hearth, Structure B05 (Göltepe, Early Bronze Age).
Plate 12: Tin x-ray map of crucible, MRN 537, cross-section (SIMS; Mieke Adriaens).
Plate 13: Microprobe image of glassy crucible accretion (Ian Steele, University of Chicago).
Plate 14: Replication of crucibles with Celaller clay (Bryan Earl).
Plate 15: Ore beneficiation, vanning with a shovel (Oriental Institute courtyard experimental smelt).
Plate 16a: Tin metal prill from experimental smelt, using Göltepe ore materials;
16b: Glassy slag envelope from which the tin metal prill was released upon grinding (experimental smelt).
Plate 17: Experimental smelt, Celaller Village using three blowpipes.
Plate 18: XPS metallic tin from powdery ore material (Göltepe, Early Bronze Age).

ACKNOWLEDGEMENTS

I am grateful for the help of the following institutions and individuals who contributed to the making of this book. The surveys in the Taurus Mountain area and the excavations at Göltepe and Kestel were conducted under the auspices of the Turkish Ministry of Culture, Directorate General of Monuments and Museums, and the Niğde Museum. Special thanks go to the following institutions for financial support of the project: the National Geographic Society, the National Endowment for the Humanities, the Institute of Aegean Prehistory, the Smithsonian Institution, Boğaziçi University, the Turkish Mineral Research and Survey Directorate, the University of Chicago Oriental Institute, Antwerp University, and Dumbarton Oaks.

The team both in the field and in the laboratories consisted of Aslihan Yener, Sylvestre Duprés, Behin Aksoy, Fazıl Açıkgöz, Bryan Earl, Hadi Özbal, Gül Pulhan, Aslı Özyar, Daryo Mizrahi, Fatma and Mehmet Karaören, Pam Vandiver, Paul Craddock, Allan Gilbert, Mark Nesbitt, Ian Steele, Mieke Adriaens, Elizabeth Friedman, Ercan Alp, Brenda Craddock, Ayşe Özkan, Tony Wilkinson, John and Peggy Sanders, Edward Sayre, Emile Joel, Jane Peterson, Jennifer Jones, Eric Klucas, Tom Chadderdon, Steven Koob, Ergun Kaptan, Cemil Bezmen, A. Minzoni-Deroche, Necip Pehlivan, Atilla Özgüneylioğlu, Lynn and Sheelagh Willies, Phil Andrews, Simon Timberlake, and John Pickin.

I am especially grateful for the mentoring, partnering and collaboration of particular individuals who made this research possible. They include the support and advice of Henry Wright, Robert McCormick Adams, William Sumner, Malcolm Wiener, Prudence Harper, Lamburtus van Zelst, Edward Sayre, Ron Bishop, Jaylan Turkkan, Carol Bier, Guillermo Algaze, Aptullah Kuran, Ergun Kaptan, and Hadi Özbal. Special thanks go to Simrit Dhesi for her careful and meticulous work in editing the volume, Jesse Casana for the computerization of the illustrations, and Elise Beyer for the final polishing of the manuscript.

CHAPTER ONE

THE RISE OF COMPLEX METAL INDUSTRIES IN
ANATOLIA, ANCIENT TURKEY

Introduction

Consider the role of metals in the complex fabric of every day life in the ancient Near East. The quantities and diversity of metal tools, weapons, personal decoration, building materials, monetary standards, and coinage are dramatic testimony of its significance. While one might focus on the manufacture of metals as products, in this book the focus is on the study of metals as a cultural, economic, and industrial process. This book will discuss the rise of multitiered, complex metal technologies in the highlands of Turkey during the fourth and third millennia B.C., and present evidence of specialized production complexes at the settlement site of Göltepe and Kestel mine located in the south-central Taurus Mountains (Fig. 1).

One of the most striking features of Anatolian metallurgy is its precociousness. The earliest occurrences of metal objects date to the Aceramic Neolithic (8th millennium B.C.), the beginning of settled farming communities and animal and plant domestication. These aceramic sites are part of the growing number of settlements associated with increasing sedentarism and the earliest-known village societies. The aceramic site of Çayönü, dated by radiocarbon to c. 7250-6750 B.C., attests to this precociousness with an astonishing 4,000 malachite and native copper artifacts. Malachite was mostly used for beads, whereas other copper metal artifacts such as pins and awls were annealed and work hardened; one object had a high trace level of arsenic, suggesting the use of native ores as natural alloys. The ductility of copper was recognized very early and the strength, range, and colors of functional alloys were discovered in the late 5th, early 4th millennium B.C.

Excavations in Turkey have revealed a population with considerable technological skills and distinct strategies for manipulating their environment. These populations are profoundly associated with the ability to exploit and organize their diverse, resource-rich terrain with unbounded inventiveness. A second discernible feature is that they were heir to a metal technology which had experienced unparalleled development since the eighth millennium B.C. While the lowlands of Turkey are fertile with considerable agricultural potential, a value-added advantage is an environment rich in metals, minerals, and wood. These resources abound in the numerous mountain ranges such as the Black Sea Pontic, Taurus, Antitaurus, and Amanus, to name only a few. Uplifted as a result of colliding tectonic plates, these mountain ranges contained massive metalliferous deposits which were easily accessible. Thus, early urban settlements in agriculturally fertile areas of Anatolia had strategic advantages over the neighboring featureless lowlands to the south, by having immediate access to metal rich deposits and forest supplies. Clearly, a large number of mountainous source areas were quickly integrated into exchange relationships, suggesting that resource procurement also played an important developmental role. By commanding rights of priority over these resources, these areas could have an economic risk strategy that would provide insurance in times of financial difficulties.

Within the geographical area of Turkey, a diversity of distinct metal-producing locales exists thereby illustrating the major steps in the processing of metal in antiquity—a technological continuum spanning mining, ore dressing, smelting, and casting. While most synthetic analyses of these prehistoric technologies take into consideration particular aspects of these processes, what is missing in these accounts is an understanding of the interaction of the parts, that is, a clearer perception of how the industries were organized as cultural and economic systems beginning with the extraction of the ore to the final fabrication of the artifact. For example, scholars seeking to locate metal sources, who are inevitably unaware of the high energy fuel requirements, often assume ores were transported over long distances from the mines to the urban sites and then ultimately made into artifacts, whereas, often twice sometimes three times the tonnage of timber or charcoal is needed per ton of ore, making it in most cases inconceivable that most ores were transported to the urban sites since the fuel needs for smelting would exceed the ores. Energy procurement, a secondary interconnected technology, thus becomes a critically important part of metal production. Likewise, when approaching the problem through the perspective of ore preparation, one finds that groundstone tools are profoundly associated with metal production technology. Battering tools, hammerstones, grinders, peckers, and bucking stones are all essential parts of metal technologies. Thus lithics become one of the more singular features of the often, mountain-bound industrial sites.

Not only was mining back-breaking labor, but the technology underlying it also involved astonishing feats of engineering. Oddly, only metals experts and mining historians have understood that extracting the ores entails a high level knowledge of material science coupled with organizational skills. Ore exploration and processing, beneficiation (preparation and enrichment), and initial rough smelting is the first production stage of a metal object. This major industrial tier is hardly mentioned in craft specialization and production theories. Depending on the mining technology used, a large number of individuals could be working at mining sites at any one time. These include miners at the ore face, ore carriers, woodcutters, charcoal makers, and smelters. Thus the production of metal represents a major investment of labor in mining regions. Producing metal in multiple complex stages, which involves more than one geographical location, processing polymetallic ores, and the utilization of skilled labor, is no mere craft; it is an industry.

In this book, a metal industry is considered to be complex when two or more production tiers are in operation—one is the mining, extractive, and rough processing technologies tier and the second, in tandem, the workshop technologies tier in the urban centers. The degree to which both are controlled by the same polity is the extent to which it can wield control over raw materials. Thus the interplay of different ores and metal processing technologies within a bounded cultural context sets the stage for new concepts of how man exploited the natural environment. As Anatolian urban centers became increasingly complex, there grew a concurrent need for innovative technology and metallurgy. These factors reinforced each other in complex ways.

Competing views are reviewed about the prevalent focus on metallurgy skewed through the eyes of the end users, the urban centers. Clearly, major technological and organizational transformations were already occurring in the highland frontiers of the Near East, well exemplified by Turkey, ancient Anatolia. The same development can be seen in neighboring metal-rich regions—an "arc of metals" including the Caucasus, Iran,[1] and the Balkans (Tallon *et al*. 1987, Renfrew 1986, Caldwell 1967, Caldwell and Shahmirzadi 1966, Chernykh 1992). This study will, however, focus on central and eastern Turkey. Precocious techniques, such as fabricating copper objects by annealing and hammering into sheet metal in the Aceramic Neolithic period and casting in the Early Chalcolithic period, speak of an already-developed technology supplying the increasingly urban local polities.

While it is tempting to point to the environment as determining the rise of metallurgy, a more complex analysis suggests that cultural factors were just as pivotal. The emerging complex states in Anatolia set the stage for metallurgical transformations although the availability of resources were not the sole determinant. Doubtless, the development of complex agricultural societies and markets in lowland areas such as Syria or Mesopotamia

---

[1] For example, at Tepe Ghabristan, dated to c. 5000 B.C., finds include crucibles, open molds (bar ingots), tuyeres, slag, 20 kilos copper ore (malachite), 2 silver buttons from Level 9, lower Level 10, a shaft hole ax, hammers, picks, and adzes—a complete tool kit for a copper smith (Majidzadeh 1976).

increased the demand for metals beyond that of decorative and prestige items. The highland areas of Anatolia capitalized on these new metallurgical needs during the critical formative periods of urbanization. Preexisting exchange systems tapped into the innovations created by the technologically critical mass which had developed in particular resource areas (Marfoe 1987). The development of metal extraction technologies is examined here, especially in the Taurus Mountain resource zone, which contains deposits of silver, copper, lead, zinc, tin, and iron—all the metals of strategic importance through antiquity. Profoundly associated with these resources are the industrial production sites such as Göltepe. Workshops in the reciprocal lowland town sites situated in these agriculturally fertile areas would have received preprocessed metal products from the resource zones. These posited lowland workshops are areas where the specialized crafts of refining the rough first-smelt metal, alloying, and then casting the molten metal into idiosyncratic shapes were located. It is worth reiterating the obvious point that exchange networks tapping into the resource areas were established in the preceding periods and were at least maintained and possibly strengthened during the Early Bronze Age.

The Intellectual Framework

For the last fifty years, unilinear evolution has been the most prevalent explanatory model used by archaeologists to explain the rise of metallurgy in southwest Asia. We can recognize a gradual increase in complexity of metal use from mineral pigments in the Palaeolithic period, to cold-hammered colorful stones (native copper), to the melting and casting of native copper, and, finally, to the manufacture of advanced alloys and iron. Thus innovation in metallurgy was considered to be the logical outcome of human developmental processes. It is usually thought that the advantages of innovations in metallurgy were so clear that any intelligent group of people would adopt them, leading to further experimentation and development. Human progress was thus often largely described in technological terms as a progression from the Old Stone Age, through the Copper Age, and finally to the Iron Age (Daniel 1967: 79-98, Thomsen 1836). Later refinements to this scheme add a Chalcolithic (metal and stone) period, an Arsenical Copper Age, and a Bronze Age (Forbes 1963, 1964a and b, Esin 1976c). Each of these temporal classifications was also subsequently divided into five substages called early, old, middle, young, and late (H. Müller-Karpe 1974).

Metal objects recovered in large urban sites in the Near East were singled out as representative of this evolution. Attempts were made to pigeonhole the finds into the predetermined technological stages, even as anomalies began to emerge with new excavations. Thus metal artifacts excavated from various burials, hoards, or urban contexts were catalogued, described, and dispensed with as yet another set of material finds. Often published as a separate chapter or appendix at the end of excavation reports, the study of metals was relegated to the enumeration of artifact types or the absence-presence of particular metals found at the site as indicators of trade. Exquisitely accurate drawings, typologies, and distribution maps of weapons, jewelry, and tools across the Near East and Europe became the mainstay of publications.

Also salient in early metals research were the emphases placed on sweeping questions of origins and concerns about the distribution of these technologies within southwestern Asia. Traditional concerns in archaeology were where metallurgy began, how it evolved, and the location of sophisticated centers of metal workshops. Some theorists suggested a unique origin of metallurgy consisting alternatively of people with the knowledge of metallurgy migrating into a region or transmission of the "idea" of civilization and technologies in an unspecified manner. This prevailing perspective also included notions about the origin of civilization at Babylon, the search for biblical origins, and locating the fount of agriculture in the fertile crescent. Childe (1936), who emphasized the importance of metallurgy in societal development, saw this technology diffusing from Mesopotamia to Anatolia to the Aegean and from there to Europe. This idea was to counter an earlier belief put forth by Frankfort (1928) locating the cradle of metals in the as-yet undefined Caucasus, a convenient area much posited as the homeland of exotic peoples and materials. The latest manifestation of these theories relocate sophisticated metallurgical techniques back to Mesopotamia from where they are said to migrate outward to less civilized areas. Thus on the study of origins for example:

> I believe that the discovery whereby a hard, intractable rock is turned into a soft, pliable, and malleable metal, was a unique discovery, not one miraculously repeated in much the same way at different times in different parts of the world...Any discussion of origins must first face the fact that such investigations are no longer popular. There is a growing feeling among anthropologists that origins are not important, that there are better things to do than attempting to determine who came first, and that such research is, more often than not, a thinly veiled cover for nationalistic puffery...I would argue, on the other hand, that origin is part and parcel of understanding...What is within our grasp is a correct understanding of beginnings insofar as they are preserved in the existing archaeological and historical record. (Muhly 1988: 3)

These theories acknowledge the very early manipulation of cuprous minerals and native copper at sites such as Aceramic Neolithic Çayönü in Turkey (Maddin, Stech, and Muhly 1991, Braidwood *et al.* 1971) and Ali Kosh in Iran (Smith 1969), which are located in proximity to ore sources. Nevertheless, the more sophisticated metal technologies are said by proponents of this group of theorists to evolve in core urban centers in the Near East and later to diffuse back to the less developed but resource-rich peripheries. This notion is so deeply ingrained that metallurgical developments visible in highland Anatolia, eastern Europe, and Iran were believed to be directly related to the movement of technologies from Mesopotamia. In terms of metals, for example, the tools and weapons of Mesopotamian manufacture, as well as their technologies, are often seen by this group as having diffused into Anatolia, even in the third millennium B.C., a period known for its technological sophistication (Childe 1951). Indeed, this idea often appears in scholarly writings about the beginnings of urbanism, i.e., the higher the socio-economic complexity, the more innovative the metallurgy. Certain complex states in Mesopotamia were designated as having the primacy of metallurgy since having raw materials did not necessarily assure growth in technologies. In fact, technologies in resource zones were said to develop later than in areas without deposits of metals (Muhly 1989: 1), resulting in locating innovative metallurgy away from the raw materials.

Not considered at all were agriculture-based complex states in lowland Anatolia located in closer proximity to the raw materials. Therefore Mesopotamia was the only complex society deemed worthy of consideration as a model of production. Understandably this prejudice was based on finds dating to the third millennium such as those from the Royal Graves at Ur and Kish (*see* Moorey 1985). But as archaeologists began to excavate in the areas surrounding Mesopotamia, extraordinary hoards of metal and metal-working workshops began to emerge. The diversity of metals found in Chalcolithic Varna (Renfrew 1986) and alloys found at Nahal Mishmar in Israel (Shalev and Northover 1993) generated new questions about multiple origins of metallurgical innovation versus the one origin model. Renfrew (1986) extended the idea of independent invention and development of copper metallurgy to southeastern Europe as well. Because the origins of metallurgical technology could not be neatly fit into categories of "originators" versus "recipients" the rise of material science and the nature of its role in the developing complex societies was not clear.

Another line of inquiry distinct from theories seeking metallurgical origins was the increasing use of instrumental analysis which tried to shed light on metal sources. From the beginning (see Junghans *et al.* 1960, 1968, 1974, Esin 1969), scientists who used instrumental techniques sourced the metals by using trace element analyses, and some applied statistics to the data. The absence or presence of index trace elements (for example, gold or cobalt to fingerprint copper ores) suggested ore sources for thousands of analyzed artifacts and thus added credence to theories about trade patterns, technological development, and the availability of resources (Hartmann 1978, Hartmann and Sangmeister 1972). Without regard to its archaeological relevance, distant mining regions were hypothesized to supply vast numbers of artifacts with often erroneous attributions (Waterbolk and Butler 1965). Voluminous lists of elemental compositions joined the often rarely consulted addendum of excavation reports. Key factors omitted in such analyses were ore composition changes within complex geological deposits and the mixing of scrap metals. Archaeological context, too, was almost always the last priority in the statistical manipulations of data. That is, metals found as parts of burial assemblages, coherent hoards, or within sealed floor deposits were evaluated as a meaningless numerical soup. It must be emphasized that the application of objective numerical techniques to archaeological data should be archaeologically sensitive (Bishop and Lange 1991, Bishop *et al.* 1990).

As interest in the relevance of metals in the socio-economic and political aspects of culture grew, scientists asked increasingly sophisticated questions about the role of metals in the processes of cultural change (Wertime 1964, 1978, 1979). For example, fresh observations on the development of metallurgy often echo the emphasis on environmental context in research on the domestication of plants and animals. In the late 60's and early 70's archaaeological interest in cultural development, change, and transformation processes received a tremendous impetus from ecologically oriented processualist research. Theorists moved away from describing artifact and architectural typologies and radically altered their investigations into understanding the factors behind the shifts from a hunting and gathering subsistence to farming and pasturalism. The domestication of plants and animals, a major "revolution" according to Childe (1936), was localized in the Near East at one of a number of oases where the propinquity of man and animals resulted in the inevitable discovery of agriculture. In the postwar years Braidwood and his colleagues (Braidwood *et al.* 1983), drawing from a diverse array of disciplines, reconstructed the palaeoenvironment and designed models of agricultural productivity in natural habitat zones. Transformations in subsistence were seen to have taken place within areas where the wild forms of domesticable plants and animals already existed. By the 1990's it was evident that when populations lived in regions of great ecological diversity such as the posited natural habitat zones, the combined effects of climatic change, population pressure, behavior, and geography were all factors in the rise of socially organized communities and agricultural economies. These associative factors are equally relevant for metallurgy as well.

Many archaeologists have regarded metal technology as peripheral to what is yet another technology, agriculture. But in this ferment of inquiry into the nature of change and agricultural process, considerations about transformations in metallurgical technology developed as well. Childe's (1944) contribution to the discussion was to integrate technology into a socioeconomic web of associations and interactions. In this view, technology was a prime mover with metallurgy playing a critical role in the increasing sophistication of societies arising in the Near East. The effect of the plow on agriculture, the development of skilled labor and craft specialization, and the effect of advanced weapons and tools all determined the rise of urbanism and civilization, according to these new formulations. That is, a competitive edge was conferred upon those who controlled the means of production. As with his models of the Neolithic Revolution, Childe considered the rise of metallurgy to have had a consistent impact on the productivity of labor. Further, he typologized the stages of this evolutionary development such that the first stage was the use of metal as ornaments. This technological stage was considered to be a continuation of stone-working methods of grinding and cold hammering native ores. In the second stage weapons and ornaments revealed new alloying, and implements were rare and adapted to exclusive industrial use. Progressing to the next stage, copper and bronze were regularly used in handicrafts but sites still yielded stone tools. And finally technology drove increasingly sophisticated organizational changes in urban settings leading to stratified societies.

Despite eloquence, passion, and controversy about technological deterministic models and linear evolutionary schemes, these remain a matter of much current debate and even continued acceptance abroad (Chernykh 1992). However, in the United States an anti-technology bias in the practice and literature of archaeology has dampened the adoption of materials science approaches to the study of metals, ceramics, glass, and other archaeological remains (De Atley and Bishop 1991, Yener 1994c). Clear-cut formulations about how the study of material science fits into anthropological inquiry are lacking, leading to a number of misconceptions about the potential scope and utility of archaeometallurgical research. Recently, out of relative obscurity, new directions (Adams 1996, Basalla 1988) have been taken by an increasing group of archaeometallurgists and a coherent system of ideas has finally emerged.

In the 1970's and 1980's regional studies of metal production and questions about the rise of craft specialization gained impetus and started to replace broader questions of origins. Beginning with site surveys and excavations (Rothenberg 1972, 1988, 1990, Lechtman 1976), later research focused on the influence of technology on the society and the environment (Lechtman 1991, Hong *et al.* 1996). Textual documentation provided important information about the transition from the Bronze Age to the Iron Age and its new technology (Hallo 1992, Brinkman 1988). An emerging group of archaeologists began to place metallurgy within a much broader socio-economic context (Pigott 1991, 1996, Killik 1991, Ehrenreich 1991) and began to question the utility of earlier models. They explored how the study of metals within a wider range of material culture contributed to anthropological inquiry (Heskel 1983, Heskel and Lamberg-Karlovsky 1980). Articles began to be published demonstrating that metal working was a complex and dynamic social enterprise, even in the most common of artifacts (Geselowitz 1988, Hosler 1994, Smith 1981, Lechtman 1988). As Godoy (1985: 199) succinctly put it, "despite his antiquity, the miner, like Geertz's peasant, was recently discovered by anthropologists."

While models of specialization and production organization were utilized and tested in a whole range of other materials such as ceramics, obsidian, and jade (Brumfield and Earl 1987, Rice 1981, 1987, 1991, Bishop *et al.* 1982), evidence was emerging from the study of metals that these cultural dynamics not only interacted with metal technology, but that they were all products of culture-related behavior and social processes (Lemonnier 1989, 1993, Epstein 1992). It is debatable whether one can ever articulate the ritual and ideology of prehistoric mining in the Near East as so expressively noted by some anthropologists working in Africa and Bolivia. In Bolivia, for example, extraordinary notions emerged such as tin ores being likened to a living substance, replenishable by Satan when periodic libations were poured (Nash 1979). Nevertheless, one can still approach the pivotal role ideology plays in metals, and, through metallurgy, the society. With a combination of metallographic and stylistic analyses and ethnographic information, archaeologists began to investigate the differences between utilitarian and "expressive" artifacts in Africa (Childes 1991). Thus metallurgical information was sought about indigenous ideologies and its reflection in the production of metal artifacts.

At the same time, an anthropology of technology was being defined (Lechtman and Steinberg 1979, Pfaffenberger 1988, 1992). These models aimed to delineate the cultural factors behind the strategies of organizing and selecting technology (Epstein 1992, Shimada 1990)—views which are the opposite of technological determinism. For example, impressive metal production and distribution systems were found in Indonesia at a village level with only minimum hierarchical ranking (Pigott and Natapintu 1988). Furthermore, while metal tools and weapons were viewed functionally, burial assemblages demonstrated that metals were important status markers, and in their exchange reinforced social connectivity (Kristiansen 1987, Helms 1993). These studies drew out the "interpenetration and dynamic interplay of social forms, cultural values and technology" (Pfaffenberger 1988: 243). In short, it was important to understand how the society worked in order to understand the impact of metals, and vice versa.

Metal Production in Highland Anatolia: Innovation at the Frontier

Another important impediment to properly defining the role of metals in southwest Asia has been the relegation of the highlands to retrograde peripheries. Canonized in world-systems models, this view places peripheral areas as suppliers of raw materials to urban centers as part of a large-scale economic system (Wallerstein 1974, 1980, 1989). In these constructs, the social and economic development of resource areas is limited. Peripheral areas are considered stagnant in terms of development, and exploited by the core region as part of inexorably asymmetrical and dependent relationships. Simply put, the raw material suppliers will develop expensive non-local tastes and habituate on finished luxury commodities from the developed societies, thereby leading to underdevelopment.

Speculations of this sort are not surprising, for it is typical today to assume that regions on the margins of areas in which complex states developed were passive receivers of innovations that derived from more sophisticated centers. However, recent investigations have resulted in redefining the interactions of these frontier zones and core urban areas as part of larger networks of relations. Aside from arguments about whether or not concepts of global economy can be applied to antiquity, major differences of opinion arise from varying definitions of these peripheries (Brumfiel and Earle 1987, Schneider 1977). Kohl (1987) has pointed out that prehistoric world systems had only minimal similarities to those of modern times, but were, nevertheless, very useful as a more encompassing perspective. In particular, he likened the ancient Near East in the third millennium B.C. to a patchwork of core regions tapping into their hinterlands often overlapping with those in proximity, and not one of a monolithic core zone and periphery. Societies in the margin of larger states would have the choice of entering into mutually beneficial relationships with differing neighboring regions. In the present context it would be the establishment of connections between Anatolia and the Aegean, Cyprus, the Black Sea coasts, and Syro-Palestine if Mesopotamian relations became too solidified and unprofitable.

Coincident with the idea of societies being open systems with respect to their neighbors (Trigger 1984) is a view of the peripheries as regions of dynamic autonomous development, growth, and innovation (Turner 1920). Described as the frontier, settlements in these zones developed independently and technologies are seen as being initially refined in peripheral areas close to natural sources. For example, in the Classical period Rome had organization, trade, order, use of money, and law, but nevertheless metallurgy was more advanced in eastern Europe and Britain (Tylecote 1976: 53, Mokyr 1990: 24-29). Some of the areas where these frontier polities are found are rugged mountains and forests. This alpine terrain falls just between the Anatolian central plateau and the Syro-Mesopotamian lowlands, areas that are now archaeologically fairly well understood as a result of past research. Strategic passes, mountain-top settlements, and fertile intermontane pockets of high agricultural yield, bolstered by an abundance of wild game, define the ecological setting of highland Anatolia (Yener 1995b). In these mountainous regions, sites such as Göltepe in the Taurus would have exhibited complex interrelationships with their reciprocal lowland centers, often providing them with semi-processed metal products. This uniquely frontier aspect of Anatolia affords us the opportunity to study indigenous developments which can be better understood when explored on their own terms.

Considerable research has concentrated on the nature and intensity of the contacts between "frontier" Anatolia and other Near Eastern centers. Long viewed as the cultural and economic periphery of "Greater Mesopotamia," or as a borderland, Anatolia is often described as a land bridge between Mesopotamia and Greece, thus hampering a clear understanding of indigenous developments. Anatolia has been consistently defined as a provider of resources such as obsidian in the Neolithic and Chalcolithic (Renfrew 1977), and minerals, ores, metal, and timber in the later Ubaid period (Oates 1993). The appearance of Uruk-related assemblages in northern Syria and eastern Turkey has been postulated to be early signs of "colonies" that capitalized on these resources (Algaze 1989, 1993, but see Stein 1990). The role of Anatolia reflected in the literature is then as a raw material provider, that is, a resource zone of particular interest for Mesopotamian concerns.

Indeed, this impression is reinforced by archaeological evidence indicating that the large-scale commercial networks of the Assyrian trading colonies linked central Anatolia with northern Syria and Mesopotamia in the subsequent early second millennium B.C. (Larsen 1987). The complex commercial strategies of this highly sophisticated silver, gold, textile, and tin trade is given voice in the over-20,000 cuneiform tablets, written by the Assyrian merchants, which have been unearthed in central Anatolia at the site of Kültepe Kanesh (Özgüç 1986). The articulation of this trade by the foreign merchants and the muteness of the local Anatolians about their own systems has added to this slanted view. Nevertheless, tantalizing glimpses can be caught in these texts of a local, often troublesome intra-Anatolia trade, which the merchants had no control over, even during this period of intense "colonization" and economic pressure. There is no doubt that a strong intra-Anatolian system existed in copper, iron, Anatolian textiles, and other commodities. In fact, excavations at the lower town of Kültepe has revealed extensive metal-working workshops where scores of molds, furnaces, and copper tools were found in great profusion (Özgüç 1955, A. Müller-Karpe 1994). Attempts to monopolize local textile trade, restrict iron trade, and penalize smuggling and tax evasion are often the topics documented in the cuneiform tablets (Larsen 1976, 1987). These references to local Anatolian socio-political configurations may reflect vestiges of dynamic political combinations that existed even prior to the colony period. Thus, we will argue against this tendency to view Anatolia as one, undifferentiated "highland" and as nothing but a resource zone for Mesopotamia, and define some of the diverse trajectories of local exploitation and development.

Case Studies of Production Models

The case studies of production models set the stage and present contextual information which will allow for a better understanding of third millennium B.C. mining/smelting complexes such as Kestel and Göltepe. A number of archaeological site case studies exemplifying various production models and organizational strategies are presented in the ensuing chapters. These are 1) the nascent specialization of metal production, storage, and distribution at Değirmentepe during the Ubaid period (4500-3900 B.C.), 2) technological changes and cultural choices at Uruk-related Arslantepe during the Chalcolithic periods VII, VIA, and subsequent VIB (c. 3800-2900 B.C.), and 3) specialized function mining and ore processing at Early Bronze Age Kestel tin mine and its contemporary miner's village, Göltepe (3000-2000 B.C.).

The first case study fits within the Ubaid-related Chalcolithic period when innovative metal technology played a major role in the culture and economies of the resource-rich highlands of Anatolia. This period is emphasized because in the transition from trinket metallurgy to the production of large-scale tools and weapons, one confronts the nature of unilinear neoevolutionary logic and its shortcomings. Embedded in the processing technology at Değirmentepe and other Ubaid-related sites are important clues to understanding the shift of emphasis from minor-scale decorative metals to the production of tools and weapons. Paralleling this is a transformation of strategic alloying technology, from the use of copper to utilizing a whole range of polymetallic ores. Provocative evidence suggests that over 30% of the Ubaid-related site of Değirmentepe in eastern Turkey is directly involved in the production, storage, or distribution of copper and other related minerals. The predominence of Mesopotamian cultural features at Değirmentepe puts into perspective the galvanizing effect that the complex metal technologies must have had in order to draw the agriculturally affluent to this site. Technologically advanced regions were a magnet not just for the raw materials, but for the accumulated technological know-how. A cadre of specialized craftsmen with an advanced knowledge of material science had developed the metallurgical expertise to produce an array of power-and prestige-laden metals.

The subsequent Uruk-related sites in eastern Turkey and agriculturally fertile northern Syria may have acted as intermediaries linking the resource and technologically advanced zones to southern Mesopotamia through the convenient transport highways of the Euphrates and Tigris rivers (British Naval Intelligence 1919, 1942, Yener 1980). The agricultural potential of northern Syrian-southeastern Turkish sites such as Leilan (Weiss 1986), Tell Brak (Oates 1993), Tell al-Judaidah (Braidwood and Braidwood 1960, Yener *et al.* 1996), and Haci Nebi Tepe (Stein 1994, Stein *et al.* 1996) provided substantial resources, forming the basis from which to pursue these new materials and techniques.

But obtaining raw materials without skilled personnel to process them is useless. Historically the solution has sometimes been to mobilize personnel to target the craftsmen for transfers of technology. Information about the physical abduction of skilled labor is rare for the Uruk period, but some textual information exists for later periods. The capture, enslaving, and desirability (Sasson 1968) of non-local craftsmen have been oft-cited goals in Mesopotamian epigraphic materials for millennia. Whether the objective was transporting, as war booty, the skilled labor of metallurgists, textile workers, ivory workers, or chariot makers, it is obvious that tribute or force in obtaining the critical resources were other avenues for procuring metals, technology, and its products (Edens 1992, references in Zaccagnini 1983). Deportations of specialized craftsmen and his siege of the city of Aratta are recited in the Sumerian poem of Lugalbanda, the King of Uruk in third millennium Mesopotamia (Wilcke 1969: 409-12). Not only are the precious stones, molds for casting, and metals taken after the siege, but the goldsmiths as well (Zaccagnini 1983). The presence of advanced metallurgists is suggested by the explosive growth of metals and metal workshops in Mesopotamia indicated by the impressive and extensive corpus of metalwork in the Early Dynastic period. An astonishing range of techniques emerges displaying all alloys of copper, precious metals, gold, silver, and electrum. All types of casting techniques and decorative skills, including granulation, cloisonné, and the use of filigree gold wire, are apparent in the objects, exemplifying the skills and taste of these urban workshops (Crawford 1991, Moorey 1982, 1994). The *simug* or smith is well on his way to being a much sought after, full-time employee of large institutions such as the temples or palaces (Limet 1960, 1972).

Legends of far-flung Mesopotamian interaction iterated in epigraphic materials speak of the attention to the northern regions. A number of well-known third and second millennium inscriptions indicate Mesopotamian knowledge in the use of geographical terms usually associated with the Anatolian environs to denote the source of their raw materials. References to interregional contact with Anatolia can be found in historical and pseudo-historical records and accounts of Mesopotamian kings which were copied by scribes through the generations, constituting the histories of their kings and their exploits (Güterbock 1969). While some seem to be direct copies of third millennium records, others are couched in fanciful, mythological language. The growth of Uruk-related sites in these zones and the later second millennium Assyrian trading colonies are tangible results of this attention. These networks in Syria, Mesopotamia, Iran, and Anatolia are postulated in one prominent view to be tied together in a loosely defined world system (Algaze 1993, Kohl 1987).

This mention of colonies and world systems leads to the second case study, namely, the Chalcolithic period site of Arslantepe near Malatya in eastern Turkey. Recent analyses of the ore, slag, and metals note changes in a range of metal technologies through time (Palmieri, Hauptmann, Hess, and Sertok 1996). Contrary to the prevalent typological neoevolutionary models of technological change progressing from copper to bronze, information from this site suggests that the fluctuating technological changes were complex, non-linear, and influenced by political, social, and economic factors. As has been demonstrated by numerous scholars (Binford 1977, Binford and Sabloff 1982), technology is complex cultural behavior and thus metal technology should not be viewed from just a historical narrative of origins and evolutionary stages. The changes in metallurgy at Arslantepe had as much to do with available resources or technological proficiency as they did with technological styles, and, as such, are archaeologically definable. As part of the discussion of the rise of complex metal industries, the site fits into a larger range of issues concerning the role of metallurgy within the dynamic of developing states.

The third case study, Göltepe and Kestel, which represent the appearance of specialized mining and metal-producing sites, dates to the Early Bronze

Age, third millennium B.C. By this period a metals craft had been transformed into a multitiered, complex, metal-producing operation with wide networks of interaction. The first defined production tier is the extraction and smelting sites in the mountains; the second tier is the workshop production centers found at urban lowland sites. Göltepe reflects the distinct strategies of the first tier of processing rough metal products. That is, local ore extracted directly from neighboring mines is ground into a powdery consistency and then smelted into rough form. To be sure, abundant forest supplies nearby played a large role in the transformation of tons of ore into transportable ingots or rough, first-smelt metal products. From modest beginnings as perhaps seasonal, opportunistic mining by transhumant pastoral nomads, the sites developed into a highly specialized and focused industry where over 5 km^2 of mineralization were mined, and ore-processing areas were integrated into a walled and protected complex.

The organizational strategies of tin mining at Kestel and production at Göltepe reflect a dynamic, productive, and distinctly Anatolian industry. The location of mineral deposits in inaccessible areas gave rise to relatively self-contained communities. This provides an unusually favorable situation for the reconstruction of technological choices made by the ancient metalworkers. At Göltepe metallurgical data was recovered in well-defined contexts reflecting various aspects of its production phases. The social and physical organization of the tin industry underwent several changes in response to resource constraints and still-elusive sociopolitical events, and smelting techniques underwent some change. But the influx of cheaper, more readily available tin from abroad during the second millennium B.C. Assyrian trading colony period failed to introduce innovations into the process. Instead, the tin industry was extinguished perhaps by a combination of the competition and a deteriorating environment (Weiss *et al.* 1993).

In the following chapters, a history of metallurgy in Anatolia will be traced, from very early in the 8th millennium B.C. to the high degree of sophistication and industrial scale attained by the 3rd millennium B.C. Chapter Two presents the salient features of a century of Anatolian archaeometallurgical research. The main technological developments are summarized, starting from the Neolithic beginnings when luxury items were prominent. The Chalcolithic period illustrates the transformations of technology occurring in the resource zones ringing the Syro-Mesopotamian basins during the Ubaid. By the Uruk period and the Early Bronze Age, the technology of prestige and power becomes the agenda for the industrial production of metals. Current knowledge about ancient Anatolian metallurgical practices has been derived from laboratory analyses of metal artifacts, survey, and excavations of workshops and graves.

Chapter Three presents the evidence from Göltepe and Kestel, which provide the excavated data used to reconstruct the chain of behaviors that led from the transport of Kestel ore to Göltepe, to the transportation of finished metal artifacts off the site. Only information relevant to tin production parameters are presented here. Full excavation reports will be published elsewhere. The sites are integrated into a regional context and can be seen in their location relative to ore, fuel, and other metal related sites. Estimates of the scale of the smelting industry and inferences about the organization of production will be advanced. Chapter Four presents analyses of crucibles, powdered materials, and ores which provided insight into the reconstructions of smelting processes and set the parameters for several modern smelting experiments. This provides much direct insight into the most fundamental metallurgical activity, the smelting of metal from its ore, and places the reconstructed techniques in their social and regional context. Chapter Five presents conclusions and suggestions for further research.

CHAPTER TWO

THE ARCHAEOLOGICAL BACKGROUND

Introduction

Although the region also produced impressive agriculture-based settlements in the Chalcolithic period, metallurgy is the main suit of Anatolian technology. Containing some of the richest ore deposits in all of the eastern Mediterranean, Turkey was among the areas in which the earliest metal industries developed. Metallurgy expanded from this area of the Near East to neighboring Mesopotamia and Syria. Styles and traditions of metalworking in the Chalcolithic period and Early Bronze Age exhibit great creativity and the products of these techniques, the metal objects themselves, display a virtuosity that often outshines other technologies. Every metallurgical technique known up to the latter part of the 19th century A.D. can be found, with the exception of casting iron and hardening steel by quenching. Not only were newer metals such as terrestrial iron more fully worked, but the full extent of metallurgical techniques were pushed to their functional limits. Taken within the context of increasingly complex cultural developments at agriculturally rich urban centers, the impetus behind a high level of commitment to metallurgy and innovative technology in Anatolia becomes more apparent.

This review of the emergence of complex metal industries in Anatolia is necessarily incomplete, considering that the periods under review encompass as much as 2400 years and cover a belt of land about 2500 kilometers long and 750 kilometers wide, from Thrace to the borders of Syria, Iraq, the Caucasus, and Iran. Such an undertaking seems absurdly ambitious; however, a large part of the archaeological contexts have already been published in English, French, or German elsewhere (de Jesus 1980, Yakar 1984, 1985, A. Müller-Karpe 1994). In this chapter the main technological developments are summarized, starting with the Aceramic Neolithic. The Chalcolithic and Early Bronze Age metal coverage will necessarily draw from a small sample and be heavily biased toward the central and eastern parts of Turkey, most relevant to the Taurus industries.

A great many projects in contiguous regions in Iraq and Iran have been halted due to present-day political situations. As a result, investigators have refocused attention on neighboring Anatolian regions and new excavations in Turkey have yielded additional information on the emergence of metal industries. Unfortunately, a number of these have been published in Turkish and in obscure journals, making them inaccessible to a large number of researchers. Some of these have been published in *Kazı Sonuçları Toplantısı Bildirileri* [The Excavation Results Symposium], an annual series which began in 1980 and is published by the Turkish Ministry of Culture, General Directorate of Monuments and Museums. Subsequent series include *Araştırma Sonuçları Toplantısı Bildirileri* [Research Projects] and *Arkeometri Sonuçları Bildirileri* [Archaeometrical Research]. A separate but relevant series of analyses from Turkish excavations are six volumes of archaeometrical research by several universities, which were published as part of the yearly Turkish Science Counsel (Tübitak) symposiums called *AKSAY* and *Arkeometri Unitesi Bilimsel Toplantı Bildirileri*, from 1979 to 1989. Some sites below have been given more extensive treatment since the information about the metals is more difficult to find. The following section traces the metal finds from excavations dating to the Neolithic period when metals were part of an assemblage of small-scale, decorative, prestige items.

The Technology of Prestige: The Aceramic and Pottery Neolithic Beginnings

While metal finds mostly made of native copper and malachite are known to be present at aceramic sites in Turkey, the quantity of the finds, as well as the magnitude of sophisticated metallurgical knowledge that underlies their fabrication, was a revelation. Metallurgical technology consists of four processes. These are (1) cold working, (2) smelting and refining (extractive operations), (3) alloying, and (4) casting, forging, drawing, joining, and surface treatment (fabrication). Weighty evidence for the first category is the widespread occurrence of predominantly cold-worked native copper and copper oxide ore in southwestern Asia, beginning in the 9th-7th millennium B.C., in the form of ornaments and luxury items. Even earlier, potentially important ores were being collected and utilized as early as the Upper Palaeolithiceriod as pigments and this continued into the village stage of life (Schmandt-Besserat 1980). Lumps of iron oxide were found in cave contexts in Beldibi and Belbaşı near Antalya, southern Turkey, dating to the 10th millennium B.C. (Bostancı 1965). One of the earliest-known examples of an object is a perforated oval pendant of perhaps malachite from Shanidar Cave in the Zagros mountains, northeastern Iraq, that had been ground into shape (Solecki 1969, uncalibrated radiocarbon 8655 B.C.). The site also yielded a skeleton with a green stain on its tibia, perhaps resulting from an oxidized object or evidence of the use of ground copper ore as a pigment. The cultural context was the transition from hunting and gathering to farming. Obsidian characterization analysis indicated that the site was supplied from Anatolian sources in the Van region in Eastern Turkey and Solecki suggests a copper source in the same area. Zawi Chemi Shanidar, an outdoor site near Shanidar Cave, yielded copper which the director, Solecki, notes was ground like a stone as in the tradition of a Mesolithic lithic technology.

Lithic technology at Epipalaeolithic and Aceramic Neolithic sites holds clues to the production of a spectrum of minerals. In the Aceramic Neolithic period hundreds of beads, pins, and other decorative items were crafted from native copper and easily workable oxides and carbonates of copper at sites situated close to rich ore deposits. The same abrading and grinding activities which were used in the production of edibles were used in parallel production in a host of minerals. Large mortars and pestles used for nutting and grinding slabs find exact parallels in metallurgical production of grinding ores. These techniques paved the way to a developing awareness of material science. Copper occurs either in a native state (99% pure) or more commonly in the form of an ore. Surface copper was undoubtedly more available in these periods than it is today, and was discovered in the form of copper nodules, still found in Ergani Maden (Tylecote 1987, Griffitts *et al.* 1972) and other parts of Turkey (Özbal 1983). Aside from polishing and hammering native copper, attractive carbonate ores were also worked into beads and other small ornaments and tools, such as awls and pins made from wire.

Some of the earliest examples of metal in Anatolia are native coppers worked like a stone. These copper artifacts are found at early food procurement sites such as Çayönü, Hallam Çemi, and Aşıklı Höyük, some of which had developed widespread patterns of local and long distance resource utilization. Complex exploitation patterns in obtaining resources for both edible and decorative items were facets of organized community activities. Substantially large public structures and plaza-like spaces in the settlements suggest a socially organized community. Monumental buildings, large-scale sculpture, and artifacts charged with symbolic meaning were expressions of public display, most strikingly obvious in aceramic Nevali Çori (H. Hauptmann 1993), Göbelki (Schmidt 1998), and Çayönü (Braidwood and Braidwood 1982, A. Özdoğan 1995: Pl. 5). The brilliantly green- and azure-colored bead necklaces and other ornaments of copper found at Çayönü and Aşıklı Höyük were assuredly conveying symbolic personal expressions echoing larger scale examples from the public sphere. At Aşıklı some metal artifacts were found as burial gifts (Esin 1995: 73) while others were scattered on the floor in the process of being worked Frangipane (1985: 215) interprets these early metal and mineral artifacts as the result of experimentation with easily accessible ore bodies and not properly metallurgical technology. Indeed, from this increasing proficiency in manipulating materials and pyrotechnology, a material science emerged that had important ramifications in strategic resource areas.

Substantial evidence of emergent material science is forthcoming from the site of Çayönü. This aceramic site was excavated jointly by Robert and Linda Braidwood of the University of Chicago in a collaborative project with Halet Çambel and Mehmet and Aslı Özdoğan of Istanbul University from 1964 onward (Çambel and Braidwood 1970: 51, Fig. 3, Braidwood *et al.* 1971, Çambel, Braidwood *et al.* 1980, A. Özdoğan 1995). It is located in a fertile highland area replete with native wildlife and flora, approximately 20 km from Ergani Maden, one of the most productive copper sources near Diyarbakır in eastern Turkey. Dating to c. 8250-6150 B.C. (A. Özdoğan 1995: 81, uncalibrated radiocarbon), the site provides important information about the transition from food collecting to fully domesticated subsistence. It is important to note that some metal examples came from levels where subsistence depended mainly on hunting wild game (A. Özdoğan 1995: 83). Its chipped stone industry is primarily flint with obsidian as a minority and a well developed groundstone tool industry which includes beads, celts, pounders, and grinders. Although its agricultural function is stressed, part of the lithic industry would have assuredly been utilized in the crafting of metal and mineral artifacts.

Worked metal was found within the settlement in all levels, and was especially abundant in the Grill Building subphase and the subsequent Cell Building subphase. Native copper was used and over 200 metal artifacts and fragments have been found. Nearly 4000 small cylindrical and pear shaped beads, pins, rings, and awls (Stech 1990: Fig. 4) were crafted from malachite, azurite, and cuprite (L. Braidwood personal communication, M. and A. Özdoğan in press). Red ochre, iron oxide, was used in burials as well. In the Cell Building subphase, malachite was worked into discs perhaps used as inlays, ground into pigment, and used for fabricating pins, hooks, and reamers (A. Özdoğan 1995: 85). The largest concentration of copper and malachite came from two areas in a single courtyard. Malachite locations were also rich in small finds such as stone and bone ornaments and small clay artifacts, and were perhaps part of intensive craft activity in these areas. Early evidence of pyrotechnology includes the natural cement floor of the Terrazzo building, heat-treated obsidians, and annealed, native copper artifacts. Of the several thousand native copper and malachite artifacts, forty were examined including awls, beads, hooks, fragments of sheet metal, and wire (Maddin, Stech, and Muhly 1991, Muhly 1989, Stech 1990). The bulk (80%) of the finds came from levels within the Intermediate subphase, between the earlier Grill subphase and the subsequent Cell Building subphase. The implication is that the utilization of copper did not develop steadily, progressively or continuously (Muhly 1989), although a fuller assessment must await the final publication.

Optical metallography provided information about manufacturing methods, while Proton Induced X-ray Emission (PIXE) and Atomic Absorption Spectrometry (AAS) were used to determine elemental composition. The copper artifacts had high purity with few contaminants. Eighteen were pure copper, averaging 99.2% Cu, 0.02% Ag, 0.13% As, 0.02% Ni, and 0.015% Fe (Muhly 1989: 5). Some sheet metal artifacts had unusually high arsenic levels at 0.42% and 0.875% (Muhly 1989: nos: 70.115 and 78.1.13, Esin 1969: no. 18431). Some researchers consider 1% arsenic in copper to be a low-grade bronze (Pernicka *et al.* 1990: 272), however, no appreciable color or physical property changes are detectable at these low levels. Arsenic does occur as an impurity in some native copper from Iran (Tylecote 1970: 289), and principal components analyses suggest that the artifact samples and the native copper ores from Ergani Maden are well matched (Maddin, Stech, and Muhly 1991 contra Muhly 1989: 8 who says they aren't). Ergani Maden may be the source of the native copper and the malachite, however, Esin (1995: 62) suggests other copper sources such as Kızıltarla nearer the site. Ore (unworked) native copper and samples from Ergani Maden were also analyzed (Özbal 1983, Esin 1969: no. 18030 with no other traces).

Two separate metallurgical industries were identified (Maddin, Stech, and Muhly 1991: 376). The first includes artifacts deriving from a lithic tradition. The malachite was ground with groundstone tools and perforated with flints, using stone-working techniques. Often cold worked and ground, ellipsoid beads and tabular pendants were made of malachite. Muhly (1989) suggests that it is easier to drill malachite than copper metal using a bow drill with a bit made of stone such as rock crystal. Metallographic cross-sections reveal that the second type of Çayönü metal-working industry consists of native copper first hammered into sheet metal and then rolled into objects (Stech 1990: Figs. 2 and 3). Tools with square cross sections are examples of this. Working hardens native copper, but also makes it brittle; heating (annealing) allows further hardening by hammering. Annealing reflects an important technological recognition of the effects of heat on metal, implying an awareness of the unique physical properties of copper. The presence of annealing twins seen in metallurgical cross-section, characteristic of re-crystallized copper indicates heating of five artifacts (Stech 1990: Fig. 6). The awl was a heavily worked, native copper object containing 0.875% As (Maddin, Stech, and Muhly 1991: Fig. 8). Hammering or rolling it into shape may have cracked the metal. However, by reheating it to about 500° C for at least several hours the strain was considerably relieved. The authors point out that it is not clear whether hammering and annealing was done just to shape the objects, or whether they were left in the hardened condition for functional reasons. For example, one hook which was left in a hardened state was a harder and better hook than an annealed one. Native copper can be cold worked to a hardness of 150 HV; annealing reduces this to 60 HV (Tylecote 1976). Other possible processes that would give annealing twins indicative of annealing, such as post depositional situations, were considered but discounted as not being possible.

Comparable evidence of early metal working such as malachite beads and hammered sheet metal fragments of native copper were recovered from Aceramic Neolithic Aşıklı Höyük, a central Anatolian site roughly contemporary with Çayönü. Preliminary work was carried out by I. Todd (1966, 1968), followed by excavations by U. Esin of Istanbul University from 1989. The site lies 25 km southeast of Aksaray on the Melendiz River (Esin 1991, 1995, Esin *et al.* 1991) and uncalibrated radiocarbon dates (8958-8400 B.P.) place it in the 8th millennium B.C. (Esin 1995: 75-76). A plethora of neighboring volcanoes supplied obsidian to a widespread array of sites from within Anatolia itself to Jericho in Palestine (Renfrew 1977, Blackman 1986). A great number of distinct obsidian sources have been characterized and are located within the regional procurement network of Aşıklı Höyük. Analyses of plant and animal remains indicate that wild game hunting and early forms of plant domestication sustained the settlement (Esin 1995: 63).

Densely packed into two neighborhoods separated by a wide pebble street, the rectilinear mudbrick houses without stone foundations are most often single roomed, although multiple-roomed buildings also exist (Esin 1995: Fig. 5). Two large and possibly public structures (Buildings HV and T) stand out in their striking use of stone foundations, larger size, and red- or yellow-painted floors. Aside from the predominant blade and scraper obsidian tool industry, bone, horn, and copper ores were worked in abundance. Necklaces, bracelets, and other artifacts of semi-precious stone and copper were found in intramural burials (Esin *et al.* 1991: 131-132, 167, Pl. 9: no. 1). The beads were made both by rolling hammered sheet native copper and by perforating whole malachite pieces after abrading them. In one case, a woman was buried with a diadem of 52 deer teeth and 7 rolled copper beads (Esin 1995: Fig. 10, Time-Life 1995: 45-57: Fig. A 22-23). Both manufacturing techniques were also observed at Çayönü.

Instrumental Neutron Activation by E. Pernicka of the Max Planck Institute in Heidelberg and Atomic Absorption Spectrometry by H. Özbalat Boğaziçi University were used to determine composition. Metallographic analysis by E. Geçkinli at Istanbul Technical University and Ü. Yalçın of the Max Planck Institute in Heidelberg was used to determine manufacturing techniques. Trace element results indicate that one sample (Esin 1995: 77: no. AH.92.105A) had high trace levels of tin (0.32%) and arsenic (0.31%), approximating some of the high trace levels of tin found in ores from Bolkardağ and Bakır Çukuru, both part of the Taurus mineralization and not far from Aşıklı. Interpretations of annealing twins on two separate beads vary, the first being indicative of natural, geological formation processes, the other a result of annealing. Esin (1995: 66) suggests that the beads were indeed annealed and hammered, which is certainly viable given similar annealing techniques apparent at contemporary sites.

With the continued focus on the extraordinarily creative aceramic period, new excavations have recovered more evidence of early metal working in Anatolia.[1] Comparable examples of early copper finds stem from the site of Hallam Çemi Tepe. Excavated since 1991 by M. Rosenberg, Hallam Çemi is located near the Batman Dam near Diyarbakır, in the foothills of Sason Dağ which is part of the eastern Taurus range. Dated by radiocarbon to the 10th-8th millennium B.C. calibrated, subsistence relied mostly on hunting and gathering, although the pig appears to have been domesticated very early. Malachite beads (Rosenberg 1994) recovered at the site are again part of a lithic technology. Examples of native coppers and malachite worked into ornaments in the Aceramic Neolithic are by no means confined to Anatolia; the same methods of abrading and hammering minerals into shape appear in Syria as well.

The more fully developed Neolithic period in Anatolia bursts with metallurgical productivity and intensification of the most varied use of metals. Çatal Hüyük, dated to the 7th-6th millennia B.C., was excavated for four seasons in 1961-1965 by J. Mellaart (1962, 1963a, 1964, 1966, 1967) and new excavations are being directed by I. Hodder (1995). The site, which consists of two mounds, is located in a dry, open valley southeast of Konya, 11 km north of Çumra. The eastern mound itself covers about 13.5 hectares. Radiocarbon dates are 6250-5400 B.C. with pottery in all levels, however, aceramic levels may exist for at least a further 7 meters below, levels which remain unexcavated. The population is estimated at several thousand with an economy based on agriculture. Skilled craft production was well developed, as the copper and lead artifacts attest, and exchange networks fed this emerging production with other exotic items from distant regions. Beads were crafted by grinding minerals and ores into beads, and pulverized ores were used for colorful pigments for

---

[1] A curious example of precociously alloyed metal, although perhaps dubious, comes from an aceramic site Suberde in southwestern Turkey. A needle was found with 8.4% Sn content in the context of a sedentary hunters' village, although the excavators questioned its find place (level X-VI dated to 6500 B.C.) because it is a high tin bronze (Bordaz 1969: 51).

wall decorations. Paintings on house walls rendered with mineral pigments such as malachite, azurite, and cinnabar (mercury) attest to the continuity and inventiveness of cold-working techniques.

The oldest metal artifacts are trinkets and ornaments, made of copper and lead (Mellaart 1967), which were found from level IX upwards.[2] Mellaart reported that visual examination suggested that native copper was hammered into sheets, cut into strips, and rolled into beads, perhaps with heat application (Mellaart 1964, Wertime 1973: 878). The subsequent level VII produced beads, pendants, finger rings, tubes, and, again, sheet copper (Mellaart 1963a: 98, 1964). Blue pigment was made by grinding copper ore such as azurite and was used to paint skeletons in levels VII and VI. Pigment from green malachite was also found on skeletons (Mellaart 1963a: 94) and beads fabricated from this mineral were utilized as funerary gifts. Equally intriguing is a fabric from a level VI burial decorated with a thin copper tituli (Mellaart 1963a: 101: E VI 25, radiocarbon dated to the 6th Millennium B.C.), making this the earliest embroidered and beaded garment. A fragment of textile in another burial revealed a thin, polished wooden peg with traces of copper oxide and sheet metal (Mellaart 1963a: 100: E VI 5).

Lead was among the earliest ores used as a luxury item, as the beads of cerussite and galena from Çatal Hüyük VI attest (Sperl 1990, Mellaart 1964: E VII, Mellaart 1967: 104, Muhly 1989: cites their context as Çatal VIA shrine 10 burial of young woman). Bulgar Maden [today, Bolkardağ] in the central Taurus is a possible source, however, recent lead isotope ratios by the author indicate that the lead stems from an as-yet unidentified source. Mellaart (1962: 52) suggests that the source of copper ore could have been Bozkır. Nearby Sızma mine is an important cinnabar (mercury) source (Sharpless 1908), where hundreds of antlers (early mining tools) were found during ore extraction in the late fifties.

There is some doubt as to whether copper smelting technology (Mellaart 1964) makes its earliest appearance with findings of slag from Çatal Hüyük (Neuninger, Pittioni, and Siegl 1964). Slag was unearthed in House E, level VIa but the debate revolves around whether this is from smelting or unintentional melting of copper caused by the burning of the house. The absence of iron silicates suggests that the slag material does not result from crucible melting or from smelting (Tylecote 1976: 5). Molten copper oxide, which was observed, would, however, need to be heated to at least 1100° C. The latest copper objects, pins and awls, come from level II and

---

[2] These materials are housed in the Konya Museum and were examined by the author in 1993. They are now in the process of being reexamined by the Hodder team (Hodder personal communication 1999).

indicate that copper was used continually throughout the Neolithic strata. The on-going new excavations will eventually settle this issue.

Further evidence of developed Neolithic metallurgy comes from phases A and B (5500 B.C.) of the excavations in the Amuq valley, located in south-central Turkey. A stone and a clay object with traces of copper on the surface (Braidwood and Braidwood 1960: 84) are evidence suggesting early metal use. Further information is forthcoming from the Late Neolithic site of Hacılar although it is not as rich in metal finds. However, the inner surface of a sherd (crucible?) with fragments of copper and traces of copper comes from levels VII and VI (Mellaart 1970: 93, 153).

Transformations in Technology and Organization in the Chalcolithic Period (c. 5500-3000 B.C.)

If any period in central and eastern Anatolia has singular distinction in the far reaching changes which occur in metal technology and its organization, it is the Chalcolithic period. Since the period covers over two thousand years, it is divided here into two stages representing two cycles of the intrusive Mesopotamian presence: the Ubaid and the Uruk periods. Metallurgy appears to be an empirical and experimental art prior to this time. Data from trace element analyses of ores and artifact types suggest that the ancient smith may have known that certain ores produce metals with different properties suited to different functions even before the Chalcolithic period. By the end of the fourth millennium B.C., however, chemical and technical problems that had stumped ancient smiths engaged in earlier decorative manipulation of metals were being resolved. Yet, the exploitation of particular ore bodies may be unrelated to the state of the smith's metallurgical knowledge, or even unrelated to a shift in the composition of the ore from one part of a vein to another. That is, cultural choices may have dictated changes in artifact composition, which in turn are linked in a socio-economic, political, and ideological web of associations and interactions.

Topographical and geographical diversity divides Anatolia into a complex mosaic of zones, each having shifting reciprocal relationships with local small sites or distant neighbors. Within this fluid matrix of socio-economic and political relationships, there is a sharp peak in the evidence for technological developments in metal production in particular areas. A nexus of emergent metal-rich sites appear in central and eastern Turkey (Frangipane and Palmieri 1987, 1989) and there is a dramatic increase in the incidence of diverse ores and experimental alloys and the appearance of smelting operations. The sites where metal objects have been found are notable in being situated in fertile, well-watered land with high agricultural potential, that is, in the Altınova, Cilicia, and the Amuq valley. What makes these sites different from contemporary northern Syrian and Mesopotamian sites is the creative tension resulting from high agricultural yields and the proximity of easily accessible and extractable ore sources (Yener *et al.* 1996). This remarkable synchroneity is an important clue to why metallurgy emerged within the context of an increasingly complicated set of interactions between the users of metals and the environment.

The archaeological record of the Early Chalcolithic period (Ubaid-Amuq phases D-E) that preceded the Late Chalcolithic complexes shows a technical proficiency in small-scale luxury items, followed by technically complex metallurgy prior to the arrival of intrusive Mesopotamian cultural features (see case studies below). A striking aspect of prehistoric metallurgy is that it is not the same in all areas, nor are changes in one direction. Full transition of metal technology from one stage to another is not necessarily a lockstep progression. A variety of technological styles existed sometimes as a party to and often completely independent of one another. This is in spite of explicit and informed interaction in materials such as pottery, seals, and architecture connoting communication, but not adoption of technological styles. Some resource areas may facilitate a transition to a different stage, while other areas may remain well adapted by maintaining traditional styles of metallurgy. Thus, while smiths of some subregions adopted the practices of specific techniques of alloying, other contemporaries retained their earlier technical styles—and not for functional or economic reasons. This has important implications in reconstructions of change and continuity, where technological styles, too, can clearly show disjunction as with architecture and ceramics.

Within this balkanized technological horizon, particular socio-economic changes are evident in a number of emerging state societies. Conflict between emerging polities is set into motion on a larger scale; this is tangibly visible in the erection of circuit walls around sites such as Mersin XVI (Garstang 1953), Hacılar (Mellaart 1965, 1970), and Değirmentepe (Esin 1989). The protracted use of stone for making utensils and weapons now has its counterparts in metal and there is fairly compelling evidence for the production of metal weapons at sites such as Arslantepe (Palmieri 1981, Frangipane 1985). However, the knowledge of how to fabricate functional tools and weapons in metal had an unexpected payoff for the sites near metal sources. Seemingly mundane and useful in appearance, now utensils and weapons also conferred wealth and social connectivity when metal became a preferred material of exchange. Thus a flat ax, of course, functioned as an ax and had use value, but it also had storable, transmutable, and exchange value (Hosler, Lechtman, and Holm 1990, Helms 1993). Greater care in its making and decorative elements now embellished prestige-laden metal tools and weapons. Stored for example in a treasury, wealth could be measured in weights of metal—regardless of the shape that was stocked.

Processes initiated in the Chalcolithic period resulted in the breaching of a dramatic economic threshold with the variety, quality, and quantity of metals manufactured as well as the power conveyed to the possessor. But what galvanized the production of metals into a scale approaching the two-tiered complexity of the Early Bronze Age was the development of an information technology, that is, the technology of record keeping, bureaucratic devices, tokens, seals, and sealings, which propelled metal industries to new heights. Management devices appear in appreciable quantities during the Ubaid period in Anatolia (Schmandt-Besserat 1992, Yener *et al.* in press) and their relevance to metal production is best exemplified at Değirmentepe (see below). In the subsequent Uruk period hollow clay balls with sealings containing geometric tokens became a prevalent administrative device. This led to writing and archives of tablets, which are the record-keeping components of the administrative and organizational know-how of Syro-Mesopotamia in third millennium B.C. These are replete with references to metal trade, standards of exchange, exchange rates, lists of metal prices, and inventories (see Moorey 1985, Muhly 1973).

Such focused and bureaucratically organized production is dependent upon demand, trade, and wealth finance. The rapidly increasing rate of interregional metal trade with and within Anatolia (Yener *et al.* 1991, Sayre *et al.* 1992) necessarily transformed the productive activities of all participating societies (Heskel 1983). An increase in the quantity of workshops producing these newly high status and prestige metals is evident. Sites such as Tell al-Judaidah, Değirmentepe, Tülintepe, Tepecik, and Norşuntepe all had evidence of in-site metal production in the 5th and 4th millennia B.C. While the metal objects from Chalcolithic sites do indeed highlight indigenous, sophisticated metallurgical skills, their very existence at this magnitude points to a hidden production technology which operated at some strength in the mountain source areas. The end result of this shift in emphasis to metals both as utilitarian and wealth objects would be the rise of sites such as the special function tin production site, Göltepe, which would have figured prominently in these developments in the subsequent third millennium B.C., the Early Bronze Age. Clearly highland Anatolia is an area that is theoretically in a position to distribute wealth both internally and externally in the form of metals, a wealth finance that is a hedge against agricultural failure.

Whether the lowland centers entered into a reciprocal exchange relationship with the polities controlling the mines, or the control of the production was direct by politically integrating the source areas, the still mute production system that arose during these millennia as a whole is still impressive. Clearly more holistic queries concerning the provisioning of centers with industrial products from the mountains need to be answered. In response to this gap in regional information, a marked increase has been recently seen in the amount of archaeometallurgical surveys and projects all over the metalliferous zones of Turkey (Caneva, Palmieri, and Sertok 1992, Kaptan 1978, 1983, 1984, 1986, 1990, de Jesus 1980, Belli 1993, Wagner *et al.* 1989 and references). Such research into the precocious and innovative technologies in the highlands has in effect brought into focus the technological know-how that emerged in this frontier zone and is briefly summarized here.

Technologically, another transformation was occurring in the selection of ores being extracted at this time. A plethora of polymetallic ores which are a characteristic of the ore bodies in Turkey were now added to the easily smeltable oxides and carbonates being exploited in the second half of the 4th millennium B.C. Similar advances and extractive technologies occured in the Balkans, and crucibles analyzed from the Gumelnitsa culture sites at Chatalka and Dolnoslav indicate co-smelting of sulfides and carbonates at temperatures estimated to be between 1100° and 1200° C (Natalja, Ginda, and Vera 1996). Glimpses are caught of similar developments in the analyses of several Chalcolithic sites in Turkey, such as Norşuntepe (Zwicker 1977, 1989, A. Hauptmann *et al.* 1993) and Değirmentepe (Kunç *et al.* 1987, Özbal 1986), located in proximity to ore sources in eastern Turkey. Over 10 kg of slag as well as 100 fragments of crucibles for smelting copper were found in EB I levels (c. 3000 B.C.) of Nevali Çori (A. Hauptmann *et al.* 1993). Interestingly, Haci Nebi Tepe, located in the Euphrates basin near Urfa where ore sources are not immediate, also yielded considerable evidence of smelting and production of polymetallic ores (Özbal, Earl, and Adriaens 1998). Four circular bowl furnaces, slag, tuyeres, and crucibles were unearthed in pre-Uruk contact levels (phase B1). Ores found at the site include a polymetallic sulfide ore of galena-bornite-sphelarite-limonite with 43% lead (Özbal 1997) as well as copper. Analyses of the ores, slag, and prill within the slag revealed high arsenic levels (0.94% highest) as well as high trace levels of nickel which parallel the ores with natural impurities used during this time in eastern Turkey. Another aspect of these complex ore bodies is the presence of potentially alloyable ore for metals. Thus smelting these polymetallic ores would result in a diversity of low-bronze alloys. Corroborating evidence is the high level of nickel, antimony, zinc, lead, and iron which all appear in the early alloys of Anatolia (Esin 1969), thereby suggesting that compositions were sometimes dictated by the make-up of the deposit. Establishing techniques to standardize these alloys is a part of the fundamental process of decision making, that is, technical choices dictated by cultural factors.

Arsenical bronzes are prevalent in the Chalcolithic period and their use continues through the third millennium B.C. (Fig. 2). Just how the early arsenical bronzes were first manufactured is still in debate (*see* Northover 1989). Whether this was done by co-smelting copper with arsenopyrites, by adding arsenic minerals such as realgar or orpiment, or by smelting arsenical copper deposits has not yet been ascertained since the source of these minerals has not been clearly defined in Turkey. Recently, however, new discoveries in north central Turkey near Merzifon have increased the validity of an arsenopyrite-arsenic mineral metallurgy (Özbal *et al*. 1999).

Towards the end of the fourth millennium B.C. tin bronzes gain precedence, at times coexisting with other alloys (Charles 1980, 1994, Craddock 1985, Yener *et al*. 1996). The process is a "long, irregular transition from a preponderant use of arsenic to preponderant use of tin, perhaps dependent upon the gradual introduction of improvements in refining techniques" (Adams 1978: 268). Within this transition, the definition of an intentional alloy is a question that is far from settled. Present criteria are more or less arbitrary. The question is complicated by the possibility that ores were selected to give the desired alloy directly, rather than by the addition of separately smelted alloying elements. That is, so called "accidental" alloys may be the product of deliberate choice. Also important is the choice of alloy for the properties desired. The small decreases in melting point and increases in hardness and tensile strength of alloys below 5% tin do not encourage the belief that these alloys were chosen for their mechanical properties alone. However, the deoxidizing properties, desirable in casting, of these alloying elements in copper (traditionally arsenic, tin, and zinc) should not be dismissed. The ease and soundness of casting are greatly improved by even low levels of these elements, and the oxides which they produce can be controlled by suitable fluxing to remove the dross. "Thus it would appear that either tin or arsenic or the two together in any ratio, could be regarded as useful additions to copper in amounts of about one percent or above by weight. Such alloys could result from either the selection of an ore containing appropriate proportions of both elements (although this is reportedly rare), from the intentional admixture of ores before smelting, or from the admixture of metals before casting" (Adams 1978: 268). By 2000 B.C., metallurgical practice was by no means at the level of a small local craft, but approached the efficiency and scale of an established industry as at Göltepe and Kestel in the Taurus mountains. The arts of smelting, melting, annealing, forging, working sheet metals, and alloying had all been mastered (Maxwell-Hyslop 1971, Franklin *et al*. 1978), and the refining of gold and silver by cupellation of lead sulfides (Prag 1978, Patterson 1971) and the use of iron were underway (Yakar 1984, 1985, de Jesus 1980, Wertime and Muhly 1980). Production models too were diverse, depending on the constraints of available local sources, or other socio-economic factors.

A. The Ubaid Period (late 5th and early 4th millennium B.C.)

During the Chalcolithic periods in discussion, a new social order is evolving in southern Mesopotamia and the Susiana plain in Iran out of which complex societies with a centralized state structure are established (Wright 1986, Johnson 1987). Increasing settlement size and often large population agglomerations become the backdrop for a number of bureaucratic innovations in management during the first phase, the Ubaid. The increases in the import and export of goods and services are incorporated into formal administrative control systems, and devices (seals, tablets, tokens) to document the traffic make their appearance. Labor-intensive public buildings and new forms of symbolic expression appear, such as distinctive buildings with niched interiors and other shared architectural features. By 4350 B.C., Ubaid culture is recognizable up and down the Tigris-Euphrates alluvium. The fine monochrome buff-painted wares (Nissen 1988) become relatively widespread in eastern Turkey (Esin 1982c) as well as in the Arabian plateau, highland Iran, and Syria (Oates 1993), although the cultural mechanisms of their distribution are not well understood. It is important to note, however, that the scale of metallurgical finds in Mesopotamia during the Ubaid period is lower than elsewhere, suggesting developmental trajectories elsewhere for this technology.

The changing organizational structures evident in Mesopotamia are more difficult to trace in Anatolia. Several factors are responsible for this dearth of information, the most important of which is the lack of horizontal exposure in comparable agricultural zones, for example Cilicia, the Amuq, and central Anatolia, pertinent to these periods. Now, however, the Amuq valley projects (Yener *et al.* in press) and the Tigris-Euphrates dam projects of Eastern Turkey (GAP) have given archaeologists the opportunity to do regional site surveys and to excavate a number of Early Chalcolithic period sites, some relating specifically to the Ubaid period, such as Değirmentepe. Glimpses can be caught of technological, political, and economic changes on both a regional and an interregional scale all along the major riverine transit highways, the Tigris and Euphrates rivers. When combined with earlier surveys and excavations in other areas of Turkey dating to these critical two millennia, shifts in the patterns of archaeological site locations and increases in site magnitude are especially evident (Mellaart 1954, 1959,

1961, 1963b, Whallon 1979, M. Özdoğan 1977). New configurations of settlements (Algaze 1993) in previously unsettled areas, the growth of fortification systems, and a florescence in pyrotechnology typify some of these changes.

Ubaid-related cultural elements in Anatolian sites include painted ceramics, tripartite architecture and administrative devices such as seals, which spread over the whole of eastern Turkey, including parts of Cilicia. These sites are culturally united with north Syria and Mesopotamia, but interpretations of this widespread and seemingly standardized cultural expression vary. Ubaid outposts are suggested at Değirmentepe near Malatya and Ubaid-related pottery is found in small percentages at Chalcolthic period sites in the Keban near Elazığ at Norşuntepe, Tepecik, and Tülintepe, as well as at other sites (Esin 1982a-c). There is compelling evidence for a procurement system for luxury goods in copper, lead, silver, and maybe gold (Oates 1993), and it is suggested here, a search for technology. Ubaid-like ceramics and architecture resembling tripartite forms show up in Mersin level XVI on the southern coast. Further afield, Ubaid-related ceramics also appear in Fraktin in the northern foothills of the central Taurus (T. Özgüç 1956), as well as at Can Hasan in west-central Anatolia,[3] although within strongly local stylistic expressions. Imported eastern wares are rarely seen within the dominant local sequences of central Anatolia, which makes it difficult to link these sites to the Tigris-Euphrates river basin. In eastern Turkey, distinct local traditions such as flint-scraped Coba wares, typical of Syro-Anatolia, can be seen together with painted Ubaid-related wares, and to the west in areas such as the Amuq, dark-faced burnished wares appear. A stylistic unity based on intrusive, non-local Ubaid painted wares shows up in Kurban Höyük, the Amuq sites (especially Tell Kurdu, Tell al-Judaidah), Arslantepe, and Coba/Sakçegözü, among others. Clearly there is Mesopotamian interest in Syro-Anatolia but the precise social and economic relationships with contemporary indigenous Chalcolithic sites in these regions as well as local developmental trajectories remain to be properly defined without a Mesopotamian bias.

A widely held view of technological change in Anatolia maintains that in the absence of Mesopotamian demands, metallurgical change would be low. But the Neolithic and Early Chalcolithic assemblages characterized by metal ornaments and trinkets change in periods prior to Mesopotamian Ubaid-like features with the appearance of larger, more functional, technologically superior tools and weapons. Copper beads, fragments of awls, and needles were found at Early Chalcolithic levels of Hacılar IIa, Ia, and Ib (5400-5200 B.C.), but in central Anatolia, the Ubaid-contemporary

---

[3] David French personal communication; the pottery is in preparation for publication.

site of Can Hasan contained the earliest evidence of metal artifact production of large-scale proportions. A solid macehead with a shaft-hole, measuring 5.3 cm x 4.32 cm, originally thought made of cast pure copper (Esin 1969: no. 17635), was found in House 3 of level IIB, dated to c. 5000 B.C. (French 1962: 33, 1963: 34). New analyses have indicated that it was produced by hammering a solid mass of copper curving it around a central shaft hole (Yalçin 1998). A bracelet and other fragments of copper were also found in graves. Directly south on the Mediterranean coast, large chisels and flat axes were found in level XVII of Mersin (see below)[4] and in eastern Turkey proportionally substantial large tools were unearthed as well. By the early 5th millennium B.C., the production of large copper-based functionally useful tools and weapons was in practice at a number of sites. A variety of ores were utilized to fabricate luxury and decorative items, while the macehead suggests larger, utilitarian utensils were also emerging.

Even in periods contemporary with Ubaid contact, metallurgical advances are apparent in regions out of proximity to Mesopotamia's interaction spheres such as central and western Anatolia (Stronach 1957 and see Ilıpınar below). The emergent metal industries and subsequent distribution systems assuredly impacted different subsystems of Anatolian society and are much more complex than the artifacts found on excavations lead us to believe. There is, however, tangible evidence of smelting and larger-scale, non-decorative, metal artifacts from Değirmentepe, and metalworking stations are found at sites situated in the Altınova valley such as Norşuntepe in eastern Turkey. Concurrently, the site of Tell Kurdu in the Amuq produced evidence of metal finds dating to Ubaid levels (Yener *et al.* in press). Ternary bronzes, which combine copper, arsenic, tin, or lead, have been recovered from Mersin, Değirmentepe, and Norşuntepe; these were perhaps experimental alloys. High zinc or nickel levels are detected in some of these which suggests experimention with polymetallic ores or impurities coming from the use of a flux. Arsenical bronze (1% or higher As) was the first widely used alloy and arsenic-rich copper objects of superior alloying attest to the exploitation of richly colored secondary sulfide ores.

In contrast to this, the Ubaid period of the Mesopotamian lowlands reveals cultural developments which are occurring on a grander scale, but with only modest evidence of metallurgy (*see* Moorey 1985: 23-24). Indeed, this is supported by the paucity of metal assemblages in Ubaid-period Mesopotamian sites. Although precocious evidence of vitrified

---

[4] An open-work ornament, dated to the 6th millennium B.C. level XXVI of Mersin (Garstang 1953: 42, 45), was unearthed. This has a questionable attribution by the excavator. It is not known whether the hesitation reflects the sophistication of the metal object or whether the context is questionable.

materials, ceramics, and metal appear early, very few metallurgical developments are evident. This must be viewed with caution since few burials have been excavated, and consequently few grave goods have been recovered, making it difficult to assess the extent or role of metals in lowland agricultural zones. Some metal artifacts and hints of the knowledge of Ubaid-period metallurgy were found at Tello, Kish, Eridu, and Ur. Ax models of clay which had splayed forms and a number of shaft-hole and double-bladed axes were discovered in burials, echoing shapes assuredly existing in metal. The reasons for the absence of metals are difficult to assess but suggestions range from the lack of necessary raw materials (Muhly 1989: 5) to changes in both organization and acquisition behavior in the Ubaid period (Wright 1986). Although trade has often been asserted as a major factor in the rise of complex political and economic forms, little of the metal of the Anatolian highlands is reaching southern Mesopotamian Ubaid sites, and as yet, none of the production industries, that is, the technical know-how. Simple transit exchange could trickle down the visible expressions of status such as copper and lead trinkets, semiprecious stones, and the raw materials essential to the elite and those in positions of power; their emerging use for expressing power and prestige are evident in the tanged spearhead of copper from Ur dated to Ubaid 3 (Woolley 1956: Pl. 30). Ubaid pre-contact metallurgy in Anatolia, in contrast, demonstrates that arsenical bronzes and other complex-ore use had already been achieved, and casting, forging, and smelting had been established by the late fifth and early fourth millennia B.C.

Case study number 1 below, Değirmentepe, demonstrates the extent to which experimentation with different properties of metals, use of arsenic for perhaps alloying, and intra-site smelting had been established in eastern Turkey during the Ubaid period.

Case Study Number 1: Değirmentepe (Malatya)

The site of Değirmentepe is located in the southern flood plain of the Euphrates in Turkey, near the Karakaya reservoir, approximately 24 km northeast of Malatya. Initially surveyed by Serdaroğlu (1977: 114 called Adagören), M. Özdoğan (1977), in a subsequent survey, identified the site as Değirmentepe. It is at 750 m above sea level and is surrounded by the Antitaurus range with an average peak altitude of 2000 m. Located on top of a hill, the mound (8-11 m high) is flat and medium sized (125 x 200 m, approximately 2.5 ha). Excavations at Değirmentepe were conducted between 1978 and 1986 by members of the Prehistory Department of Istanbul University under the direction of U. Esin, assisted by G. Arsebük and S. Harmankaya. It was part of a broader salvage project spurred by the construction of the Karakaya Dam and was subsequently flooded in 1987.

A total of some 1000 m² had been exposed before work was terminated, revealing a significant Late Chalcolithic settlement (Esin 1981b and c, Esin 1989: Pl. 31: 1).

A total of 11 occupation levels were excavated with an 8 m-thick deposit dating to the Chalcolithic period. Levels 9-11 (earliest on virgin soil) are earlier Chalcolithic and levels 9-6 are Ubaid-related, level 4 is Iron Age, and levels 5, 2, and 1 are mixed (EB I, MB I, Chalcolithic, Iron, and Late Roman/Byzantine) due to periodic flooding of the Euphrates (Esin and Harmankaya 1988). One of these flood breaks divides levels 8 and 9. Radiocarbon dates for level 7 give a range of 4166 ± 170 B.C., which is roughly comparable to the Ubaid 3 period. The architecture, stamp seals, and ceramics are stylistically similar to Amuq phases E and F, while some ceramics suggest an earlier Amuq phase D date as well.

All areas of excavation produced a consistent range of finds: hearth/natural draft furnaces, slag, ore, pigment, groundstone tools, and utilitarian sherds as well as non-local Ubaid-inspired ceramics. Level 7 is the best-preserved and provided evidence for architectural organization of the metallurgical process. Slag, metal, furnaces, and grinding equipment were found in densely packed, tripartite architectural complexes built of mudbrick without stone foundations. These structures are notable for their southern Mesopotamian and Syrian Ubaid-like affinities, which include a large central room with altar-like table, offering pits with burned animal bones, and administrative recording devices such as sealings and seals. Atypical of Ubaid settlements elsewhere are metal-related debris in over 30% of the 100 rooms reported so far, some with quantities of slag (Fig. 3). At least one hearth/furnace was found in each architectural complex, and most are associated with metallurgical debris (Esin and Harmankaya 1988: 94, Fig. 9, M. Müller-Karpe 1993). These hearth/furnaces are found in the central room or in a surrounding magazine room. Larger, furnace-like pyrotechnological features are found in what appear to be workshops in the rooms that fill the gaps between these tripartite building complexes as well. Esin (1984: 78) suggests that rooms primarily in sectors 17H, 17I, and 16J may have had more domestic functions as apparent from the utilitarian ceramics. They also appear to be utilized for storage as indicated by the sealings and had relatively fewer metallurgical finds.

Fourteen or so buildings have been excavated and twice as many more may have existed. The characteristic tripartitite building complexes often contained a large central room with two opposing series of magazine rooms, and a staircase indicating another story. Walls were preserved to three meters, revealing doors, windows, and ceilings. The building complexes appear to be arranged in radial rows possibly facing a central public space in the center of the mound. The crowded settlement is surrounded by a thick fortification wall with bastions and recesses, which was partially excavated on the northeast, southwest, and southeast slopes of the mound (Esin 1989: 136-7, Esin and Harmankaya 1988: Fig. 2). The walls and square-sectioned post-holes indicate that a superstructure of wood topped the wall (Room EN eastern wall, Building EL).

The site has primal importance in the precociousness of its metallurgical industry and the degree to which it is specialized in this activity. Instrumental analysis suggests that sulfide ores were possibly smelted and crucibles were utilized to melt and/or smelt copper. Some use of arsenic-rich ores is indicated by the composition of some processing by-products such as slag. Metallurgy-related debris was distributed throughout the entire settlement. Not only were natural draft furnaces, slag, and metal found throughout the site, but an administrative recording system for storage and exchange of materials was fully developed as well. The smaller units, which contained ovens and quantities of copper slag, often produced groundstone tools and sealing devices. Some of these assuredly were technologies related to metal processing and its administration as well.

The organization of metal production described below was inferred from the distribution of slag and copper ores in association with architectural and household features. Information about find place was pieced together from excavation codes published in the instrumental analysis of metallurgical remains (Lyon 1997). Descriptions were also taken from the text of the various excavation reports. Quantitative exactitude and completeness is limited in this assessment due to the lack of final publication. However, an attempt will be made here to localize the metallurgical data.

Distribution of Metal-related Activities
Building I in squares 17-18F in the southwestern sector yielded the most complete repertoire of material associated with metallurgical activities. Along with its metallurgical functions, several rooms indicate that it may have been a public building with symbolic functions as well. The excavators suggest that a number of Mesopotamian religious features may indicate that the structure served as a temple (Esin and Harmankaya 1988: 92-93). These are the paintings, altar tables, and monumental hearths (Esin 1990: 48), with a pit nearby containing ash and burnt soil mixed with burned animal bones, pots containing the skeleton of children, and grain bins.

This possibly public building complex consists of a large central room measuring 7.4 x 3.4 m (total building ca. 190 m^2), surrounded by a cluster of small rooms with additional storerooms to the north. The walls of Central Court I were coated with a layer of white plaster and painted with schematic sun and tree motifs (Esin and Harmankaya 1988: Fig. 20, Esin 1983a: Fig. 4, Pl. 35: no. 3). A pair of black and red lines framed red, orange, and dark brown dots. Other painted dots and borders were found on both sides of the doorway in DU (Esin and Harmankaya 1987: 107) and wall of BI. Three types of iron ore in the form of ochre were found on the floor of AD and I, and were used in the painted decorations. The pigments are various forms of iron ore such as ochre and limonite and Esin suggests that these may have been by-products of copper production.

Several fragments of copper ore from Central Room I and BI to the south were analyzed (Esin 1986). Slag was distributed in every room of Building I: Magazines AD, AC, R, AG, K, BK, Y, and U, and Central Room I. Hearth/natural draft furnaces were utilized to produce copper as inferred by their association with slag and/or other materials. A large, horseshoe-shaped installation was located in Central Room I (Esin 1983a: Fig. 8, Fig. 13: no. 146), which contained one of the few fragments of copper metal found at the site. The hearth/natural draft furnace measures 1.25 m in diameter and 40 cm deep with a pit nearby containing ash and burnt soil mixed with burned animal bones. One sample from pit no. 323 in Room BI (Esin 1986: 155: Table 1 no. 19, identified as ore), when being prepared for analysis, yielded a copper metal prill (globule shaped), no. 19B (Özbal 1986). This indicates that the sample was actually slag, and that the metal prill inside was a product of a smelt. Another fragment of metal was found in the magazine room. Esin (1986: 145) has suggested that these prills were in actuality the ingots used for the final fabrication of copper objects. This is not unlike the copper production industry found at Chalcolithic Timna (Rothenberg 1990) and much later periods in Peru (Shimada and Merkel 1991). Pit no. 323 in Room BI yielded other slag and ash remains, while two copper slag samples were taken from the accretion inside a crucible (Esin 1986: 146). These samples suggest that the use of a crucible for melting or smelting played an important role in the copper industry of Değirmentepe. The new light shed on crucible smelting functions is discussed below.

Significant amounts of slag were also found in adjacent Rooms AU, DE, AV, and DU to the east, and section H of Room BI just to the south. Large Room DU contained nodules of iron minerals, one example in the hollow of the altar (Esin 1990: Fig. 5). Room AL yielded a domed hearth/furnace that was laid with Ubaid sherds (Esin and Harmankaya 1986: 59). Another, earlier furnace was found directly below as well. Slag and hearth/furnaces were found, but in more restricted numbers, in the tripartite building complexes and adjacent units flanking Building I. Room FC and the hearth in Room ET of Building FC to the west also produced copper slag (Esin and Harmankaya 1988: 102). Room FC also had evidence of wall paintings and nodules of iron ore used as pigment. Central Room GK of Building GK to the east contained quantities of slag and nodules of red ochre.

Perhaps profoundly associated with metal finds are various administrative activities which were suggested by the sealings and seals found in these units. Although all the units of Building I contained metal residues, Central Room I and Magazine Room AF also contained both seals and sealings. Other adjoining rooms, L, AU, DV, DY, DU, CE, BY, DN, yielded seals or sealings, sometimes both. More often than not, sealings were found in the same rooms as slag or furnaces. Central Room GK and Magazine Rooms FC+GE, BC+DS, and EE+EB contained groundstone tools, pounders, hammers, or grinders and some were restricted to the Magazine Rooms EK, FS, CT. Thus some of the slag uncovered in the rooms may be from another phase of production, that is, the crushing of the slag to release the copper metal prills entrapped in it. Some rooms with groundstone tools possibly functioned as preliminary crushing zones for further smelting and grinding of pigments. Room GK had evidence of paintings on the wall, although badly preserved (Esin and Harmankaya 1988: 96). Central Room FC and Rooms GK-DE also contained seals and sealings. Copper production and its management clearly appear to be major functions of Building I and its surrounding structures. It is important to emphasize that the production of copper is taking place within a structure suggestive of a strong symbolic context.

The distribution of slag in Magazine Rooms DH, BC, DI, and BD of Building BC to the north of Building Complex I marks this building as another copper production location. Several hearth/natural draft furnaces were found in association with copper residues in Rooms DH, DS-DT, and BM. One large furnace (Esin and Harmankaya 1986: Pl. 6: 2-3, no. 509) was built into a wall separating Rooms DS and DT. The installation measures 60 x 56 cm in area and is 55 cm deep, with an opening located to the north and a thin, deep channel reaching the north wall of Room DS. Esin suggests that it may have functioned as a duct for natural draft. Numerous hammerstones were found scattered inside Room DS. A large stone anvil presumably used to grind ore or slag was in front of the south wall and behind the oven in Room DT. Another oval-shaped pyrotechnological installation was located in room DH of the same building and was also used for copper production as indicated by the large quantities of slag (Esin and Harmankaya 1986: 60-61, Pl. 5). Larger than the first, measuring 1 m in diameter and 45-55 cm deep, a furnace dating to an earlier phase was found directly beneath it (Esin and Harmankaya 1987: 112). It has an opening toward the north and a trough heading E-W from nearby Pit 504. The archaeological section suggests that the pit and channel were connected. The furnace and the pit were clay-lined and quantities of copper slag were found in both. In the adjacent Room BM, a dome-shaped natural draft furnace was found with slag just inside the opening (Esin and Arsebük 1983: 76, Fig. 8). The association with copper slag suggests that it too was used in the copper melting/smelting operation.

Sealings were found inside the magazine rooms which also contained furnaces and metallurgical debris. These recording devices were especially prevalent in Central Room BC which had both seals and sealings, while Magazine Rooms DC, BO, DH, BM, and FB had seals or sealings but not both. Slag is reported to be less abundant in the eastern structures. However, Magazine Room CC yielded not only slag but ore as well and adjacent Rooms CF and DB also yielded slag. An oven was found adjacent to Room CV with large quantities of slag (Esin 1985a: 16). Other metal related material was found in adjacent Rooms DA and CT. A wall painting depicting a sun framed by a dark rectangular border was also found in the central room of Building EE (Esin and Harmankaya 1988). An orange-painted podium stood in the center of the room. Seals and sealings were discovered in Central Room EE, Magazine Room EB, and adjacent Rooms DO, DB, and CF. Room EL and area EU also yielded sealings. The partially exposed buildings in the northern sector also had evidence of copper working. Slag was reported from Room AS and a hearth/natural draft furnace (no. 577) outside Room EZ to the north dated possibly from earlier level 8 (Esin and Harmankaya 1987: 103). Set in a rectangular frame, it measures 80 cm wide, 1.10 m long, and about 1 m deep on the inside. The furnace was probably lit from the top through the dome. The oven-pit was filled with carbonized plant remains and charcoal as well as slag fragments (Esin and Harmankaya 1987: 103, Figs. 3 and 4). This may indicate either that multipurpose hearths were used for food as well as for working copper, or that dung cakes were used as fuel. The rooms with the pyrotechnological installations also contained groundstone tools and various types of utilitarian ceramics, such as flint-scraped Coba and dark-faced burnished wares (Esin and Harmankaya 1986: 60-61). This lends support to the suggestion that these are cottage industries and not centrally controlled workshops. Three more hearths were located in this area but no slag is mentioned. Although the presence of seals and sealings was less in evidence in these northern exposures, nevertheless, seals were found near HG and Room HB. The distribution of recording devices throughout the settlement suggests some sort of organized management, perhaps centralizing the cottage industries in its focus.

Copper Metallurgy
The metallurgical finds (ore, metal, and slag) from Değirmentepe were analyzed by Özbal at Boğaziçi University, Istanbul, and by Kunç and co-workers at Fırat University, Elazığ, utilizing various analytical techniques.

These include wet chemical methods, infrared spectroscopy, atomic absorption spectroscopy, x-ray diffraction, and optical microscopic methods. Since the greater proportion of the metallurgical materials were slag-like in nature, attempts were made to characterize the nature of these post-smelting residues to determine their technological styles. Metallurgical techniques utilized at the time were determined, such as the range of temperatures achieved in the furnaces. Furthermore, the residues indicated whether more complex technologies, such as the addition of fluxing agents to achieve the smelt, use of sulfide ores, or use of alloying materials, were part of the metallurgical repertoire. The presence of crucibles, too, suggested that perhaps some of the hearth/natural draft furnaces were used for dual-purpose functions, domestic and metallurgical, leaving the crucibles for smelting or melting. These results were instrumental in determining the types of ore used in the smelt, as well as providing clues to the sources of the minerals. Slag from a furnace operation often has a high iron content and between 1-5% copper. Crucible slag often has low iron content from 0.1-5%, similar to the slag analyses at Değirmentepe, suggesting that copper was produced by crucible smelting.

A group of 17 slag, 1 copper ore, 5 iron ore, and 2 metal fragments were analyzed by Özbal (1986) for elemental composition. He lists the iron ore samples (nos. 2, 4, 6, 7, 13) which were presumably used for pigment or as a flux in smelting (Özbal 1986: Table 2). One copper ore fragment (no. 1) contained a high amount of iron (4.64%). Another, thought to be ore (no. 19), contained a metal prill globule inside (no. 19B) and thus must be redefined as a slag, the product of a smelt. This slag contained 31.7% Cu and 0.46% Fe (Özbal 1986: 108: Table 1), a not very efficient smelt since too much copper remains in the slag thus leading to the misidentification. The metal prill was relatively pure copper and contained 98.2% Cu, 0.41% Fe, and 1.43% Sb. The second fragment of metal (no. 3) had 47.3% copper and few other trace elements. Two further samples of copper ore, cuprite (Cu_2O), were reported (Kunç et al. 1987: Table 2 and 3: nos. 17, 27) with 2% and 3% copper content, respectively. However, sample number 17 contained 7% magnesium, is reported to contain sulfur, and is also listed as part of a slag group on Table 4 since it contained nefelin and quartz.

Slag accretion taken from the inner surface of crucibles yielded 2% Fe and little else (Özbal 1986: 110, Table 3: samples nos. 16 and 22), conforming to the definition of crucible slag. The other slag samples (nos. 10, 12, 14, 15, 17, 18, 20, 23) have similar results with iron at about 3.09%. Some slag samples (nos. 8, 9, 11, 21) appear to be vitrified material with no metallic content and may have been vitrified hearth/natural draft furnace residues used during crucible melting/smelting operations.

Analysis of six other slag samples confirm low iron contents at 2-5% (Kunç and Çukur 1988a: Table 4). The low level of iron suggests that the original ores smelted were copper oxide and carbonates like cuprite and malachite (Özbal 1986: 111). If the ore had been chalcopyrite, 30-50% iron would have been expected (Tylecote and Boydell 1978, Tylecote, Ghaznavi, and Boydell 1977), although analysis of one sample does indicate the presence of chalcopyrite (Kunç et al. 1987: Table 3: no. 28).

The mineral compositions of eight Chalcolithic slag samples were analyzed to determine temperatures achieved during the smelting process (Kunç et al. 1984: Table 1: nos. 1-7, 9). All samples contained diopsite, four samples had pseudo-wollastonite, and five had quartz minerals. Calcite was found to be less than 1% and iron between 1.1-3.6%. The presence of certain minerals indicates the temperatures attained and were typologically grouped. On the basis of these components, the authors conclude that a calcitic fluxing agent may have been used and that the temperature of the furnace would have been more than 1100° C. This was confirmed by using X-ray diffraction to determine the crystalline structure of 15 more Chalcolithic slag samples and the temperature, attained a maximum of 1245° C (Kunç et al. 1986: Table 1).

This typological grouping and maximum temperatures achieved was confirmed in a subsequent article in which 11 new Chalcolithic samples were added (Kunç et al. 1987: Table 4 with slightly altered groupings). Elemental analyses of these and 14 more Chalcolithic samples were obtained and confirmed the presence of cuprite and malachite ores. The copper minerals present in all samples were cuprite and malachite with the exception of sample no. 19, $CuFe_2$, a pyrite, possibly a chalcopyrite, which contained 4.8% Fe and 1.2% Mg (Kunç et al. 1987: Table 2: same as sample no. 28 in the 1986 article); the authors suggest it may not be slag. The presence of sulfur in a number of samples suggests that the smelting of sulfide ores was achieved even in these early periods. A sample of a sulfide ore, chalcopyrite ($CuFeS_2$), found in the subsequent Early Bronze Age levels suggests that the extraction and smelting of this type of ore was continued into later periods. Sample number 18 with 7% Mg and 2.9% Fe may not be slag, but may have been used for fluxing the ore during smelting. It is important to note the relatively high arsenic (0.67-2.33%) and low iron contents which occur in three slag fragments (Kunç and Çukur 1988a: nos. 1, 2, 5). Since arsenic partitions into the metal as well as the slag, a minimum of 1-2% arsenic would have been present in the prill produced by this smelt as well, a fair arsenical bronze. Arsenic was sought but not detected in some earlier analyses (Özbal 1986) and not looked for in others (Kunç et al. 1987). Thus it is possible that arsenic content may be more prevalent than apparent. Since 1-2% arsenic content in a metal constitutes an alloy, one could argue that these analyses point to experimental, early alloying technologies.

These metallurgical installations are similar to other Chalcolithic examples where copper production was found. There are on-site natural draft furnaces and production quarters within the settlement of Tepecik (Esin 1982a: 109, Pl. 62/2-3), and smelting pits and furnaces in Norşuntepe Chalcolithic levels 9 and 10 (H. Hauptmann 1982: 50, 58-9) and Early Bronze Age I levels 21, 24, and 25 (H. Hauptmann 1982: 21, 23, 29-30, Pl. 18/3-4, Pl. 20/3-4). A similar situation existed at neighboring Tülintepe which yielded stone crucibles and slag. Tepecik also contained a Late Chalcolithic oven of a different type (Esin 1984: 102, Fig. 20, 21, 1976a: 221, Pl. 1/b).

Production and Distribution Organization
The preponderance of metallurgy-related activities in the buildings indicates that Değirmentepe was a special function site and metallurgy was its production priority. However, this does not mitigate against the existence of subsistence-related activities, as the household assemblages indicate. A number of activities were occurring concurrently in these units, as evidenced by such objects as groundstone tools, flat-axes, and shaft-hole hammers which comprised the majority of the artifacts. The assemblage also yielded stone beads, clay and marble stoppers, clay straight nails, flint and obsidian borers and engravers, and bone implements such as awls, needles, and loom shuttles. The distribution of tools such as chipped stone and bone tools, groundstone tools, and weaving tools was relatively even throughout the site. Magazine Rooms ER, ET, FV, FS, flanking central rooms, and open courtyards BH and EU yielded chipped stone tools. Workshop Room BH contained flint engravers for the production of stamp seals (Arsebük 1986, Esin and Harmankaya 1988: 100). Some bone tools, such as awls and one possible shuttle, may have been used for manufacturing textiles. Although a few possible blades and projectile points suggest use in hunting and gathering (Esin 1983b: 21-3, Esin and Harmankaya 1988: fig. 39 no. 4), the site does have domesticated cereals and plant remains (Esin 1983c: 148-9), with good soil for agriculture (Esin 1985a: 17 reported by Kapur).

One of the questions needing clarification is the organization of the industry in relation to the mines, where the fuel in the form of timber is plentiful, that is, whether the industry was complex enough to have specialized processing sites near the mines where rough smelting of newly extracted ores is achieved. New archaeometallurgical surveys by the Arslantepe team in this region has uncovered a large number of mining and smelting sites (Palmieri, Sertok, and Chernykh 1993a and b). If these sites can be accurately dated, then the mounds of slag should be iron-rich matte or converter slag which is generally found at industrial smelting sites near the mines. The slag found at Değirmentepe, however, represents a smaller-scale oxide smelting process possibly done with crucibles or simple natural draft furnaces that would not leave the large quantities of slag found at first smelt sites. This type of smelting could easily be achieved in the natural draft furnaces found in the tripartite building complexes, using ceramic crucibles and blowpipes. The small quantities of more complex ores could have been easily co-smelted in a crucible as well.

A second view of a production model put forth for the site is based on the crucible analyses. A number of crucibles were found with copper accretion and these are said to have been used in a final refinement phase, that is, the melting stage (Kunç and Çukur 1988a: 100). On the basis of this reconstruction, the Değirmentepe industry would then be a secondary one, with the original product being first smelted in the mining areas and then brought into the settlement for further working. This is certainly possible, although smelting malachite and cuprite in a crucible would not leave heaps of slag anyway. The fact that ore is rarely found at the site supports this suggestion. It is entirely possible that slag which was smelted elsewhere with copper prills intact could have been transported to the site for further refinement.

Parallels for furnace smelting during the Chalcolithic period appear at Timna, Israel where furnaces were built with local sandy clay, tempered with crushed slag. Shaped like a shaft or steep-sided cone, air was supplied by bellows and fed into the furnace through clay, tube-like tuyeres. Furnaces measured 1 m tall, with a 30 cm diameter since small furnaces could easily supply enough air to maintain temperatures. Charcoal fuel was the source of the carbon monoxide used to reduce the ore to metal. Research at Fenan, Jordan (A. Hauptmann 1995, A. Hauptmann *et al.* 1989) also provides evidence of smelting oxides of copper in crucibles. Globular prills of copper metal are produced with relatively low temperatures just above the melting point of copper. Very little slag is produced because no fluxes are used, which accounts for its absence.

The organization of copper production at Değirmentepe can be deduced from the distribution of slag and hearth/natural draft furnace installations throughout the site. Metallurgical debris and installations are evenly distributed throughout the architectural units. As noted before, all buildings contained signs of metallurgy or its storage. Although the publications are only preliminary, the distribution patterns of other tools and activity areas suggest that the distribution of metal debris is significant. There are limited quantities of slag and copper localized within the site, leading some analysts to hypothesize that copper production at Değirmentepe involved on-site copper oxide smelting. This process would then use the small copper metal prills produced for local purposes or bag and export them elsewhere. The varying levels of copper content within slags and marked variability in the raw materials utilized suggest early stages of copper oxide production. A similar inference can be drawn from the prills entrapped inside the slag. If the technology were more advanced the slag would separate out allowing the copper to be tapped. Analysis of the product, copper, and tests of the metallurgical debris do, however, indicate that a certain amount of experimentation with polymetallic ores was also taking place.

The scale of copper production is difficult to determine. If it was restricted to one area of the settlement, then some degree of intra-site specialization could be inferred. However, the only evident localization is stamp seal production (Arsebük 1986), which suggests the organizational structure of storage and exchange. Various administrative activities are suggested by the sealing practices. Over 200 stone stamp seals and bullae document a regional use of seals for the marking of merchandise or property (Esin and Harmankaya 1986: 83). Tokens were also found, further substantiating the range of management devices (Esin and Harmankaya 1986: 83: nos. D. 84-69). Stamp seals and sealings or bullae were quite common in the magazines, rooms, and central rooms of the large building complexes. String and basket impressions are visible on the reverse of the bullae. The seals were used to seal containers, jars, reed baskets, and leather sacks. An imprint of a wooden peg or nail suggests the securing of a door (Esin 1985b: 255: Pl. 2: 15). Stamp seals generally occur in central rooms (e.g., CF, BC, I, DU, GK, EE) while sealings were also found in magazines. Nineteen published seal impressions were found in Central Room DE-GK and Rooms DC-7, BD-6, DH-4, and DN-1 in Building BC. Some sealings could be matched with seals from the site (Esin 1989: footnote 36, Esin 1983a: 189, Fig. 9: nos. 1-4, Pl. 36: nos. 1-3). These local styles also appeared in the painted wall decorations (Esin 1990). Local styles could be easily discerned which suggests local exchange systems.

Other sealings were not matched and were found with imported, off-site goods (Esin 1989: footnote 38). Multiple stamp seal impressions, some with the same motif in different sizes, and some with minor changes in composition, sealed another series of materials (Esin 1985b: 255). Similarities in motifs on the stamp seals and sealings between Değirmentepe and Gawra, northern Mesopotamia, and Iran hint at a shared iconographic style. Leaf and quadruped patterns resemble similar seals from Gawra XIII-XI (Tobler 1950: 126-90). The simple geometric seal designs parallel those from the intermontane hilly and piedmont zones, such as Tepe Sialk III, Tepe Giyan, Susa, and perhaps Tello (Esin 1985b:

254-6). A developed regional, if not interregional, administrative system is suggested by the quantities of sealing devices. The number of seal impressions demonstrates a degree of administrative control of imported or exported items, however, the economic functions were at the household levels.

Other non-local items found at Değirmentepe suggest that it was part of a late Ubaid interaction sphere. The intrusive ceramics are echoed in a number of other sites and are part of the widespread appearance of Ubaid and Ubaid-related materials. Flint-scraped Coba bowls, light wares and monochrome wares, and dark-faced and red burnished wares were the most common ceramics (Esin and Harmankaya 1987: 103). The closest parallels are found at contemporary Arslantepe VII and Amuq E-F. Painted Ubaid-related decoration is rare, and a fine ware with a red polished or gray slip and a dark-color-slipped cooking ware form the rest of the ceramic assemblage (Esin and Harmankaya 1986: 64). X-ray fluorescence and instrumental neutron activation at the Middle Eastern Technical University in Ankara were used to determine the trace elements of 126 sampled ceramics (Esin, Birgül, and Yaffe 1985: 57). The Ubaid-related plain wares and the flint-scraped Coba wares clustered together in one source group, B, and thus were made from the same clay source. The red-black burnished wares were significantly different and belonged to another group, A. A third group, D, consisted of some Ubaid-related wares and one red-black burnished ware. Thus, the organization of production at Değirmentepe may be typified as independent, nucleated workshop production. Metal production, especially arsenical copper alloys, at Değirmentepe may have played a major role in the spread of Ubaid-related pottery and sealings.

B. The Technology of Prestige and Power: The Uruk Contact (c. 3400-2900 B.C.)

During the latter part of the Chalcolithic period some of the earlier trends that gave rise to state societies in the ancient Near East were consolidated. Thus by the end of the later Uruk period (c. 3500 B.C.), the formation of the earliest known cities was accompanied by the foundation of settlements with intrusive Uruk-related features in Syria, Anatolia, and Iran. As with the postulated Ubaid outposts, these would have functioned to obtain resources and advanced technology in demand in Mesopotamia. Frangipane(1993a: 159) believes that the Uruk colonial event is limited in Anatolia and considers only some settlements along the Tigris and Euphrates to be intrusive. Uruk-related elements came into contact with a "very well structured and solid local territorial organization, which seems to appear in the late Ubaid along with more 'industrialized' pottery production, and the emergence of elites." An increasingly powerful elite is reflected in the emergence of more substantial public structures, hierarchical burials, precious metals, and the adoption of various ritual symbols. Nissen (1988) notes an increase in the instance of crafts which were divided into distinct tasks by different people. This is most notable in the manufacture of pottery and seals. In the Karababa Dam area seven (perhaps 12) out of twenty level VII period settlements had Uruk-related wares (Palmieri 1985: 204). Dynamics set into motion in Anatolia at this time include the continued growth of site size and quantity in particular areas (M. Özdoğan 1977, Whallon 1979, Wilkinson 1994). This growth is evident in fertile agricultural basins such as Cilicia, the Amuq (Braidwood 1937, Wilkinson 1998), and the Tigris-Euphrates basins (Algaze *et al.* 1992, 1994).

Bronze working (both arsenical and tin) developed into an important industry in Anatolia during this period. The exact threshold of complex industrial production is difficult to pin down in the Chalcolithic period continuum, although the production of large-scale artifacts and distinct regional technological styles were apparent by the Uruk period. The widespread arsenical alloying of copper in the 4th millennium B.C. can be seen at a number of sites. One western site, Ilıpınar, yielded twenty objects of which seven were from graves and two were from excavated strata dating to Period IV (calibrated radiocarbon at 3650 B.C.). These were subjected to neutron activation and lead isotope analysis (Begemann *et al.* 1994). Arsenic contents range from 7.42% to 1.41%, which suggests intentional alloying of copper with arsenic or mixing arsenic-rich ores with copper before smelting. This may be indicated in the positive correlation of silver and gold elements (they co-occur) and in the negative correlations of silver to arsenic or gold to arsenic (they do not come with the arsenic). The authors however suggest that the "fairly constant arsenic contents are fortuitous and reflect a fairly constant composition of the ores utilized" (Begemann *et al.* 1994: 205), implying that they were tapping into arsenical copper ores. One of the lead isotope group of Late Chalcolithic objects shows consistency with ores from Serçeörenköy in Çatal Dağı in northwestern Turkey, 60 km SW of Ilıpınar. However, no arsenic-rich ores have been found as yet from this source. Advanced metallurgy and specialized skills are indicated by the widespread smelting of complex sulfide ores (Wertime 1964, 1973, Frangipane 1985: 216). Analysis of copper artifacts from Late Chalcolithic sites in Anatolia demonstrates that a low arsenic content (2-2.5%) was commonplace, and that in the mid 4th through the late 3rd millennium a bimodal distribution with 2.5-3% and 1-2% arsenic was the most ubiquitous alloy (Frangipane 1985).

Along the Black Sea coast lies the site of Ikiztepe originally excavated by U. Bahadır and Handan Alkım and subsequently by Ö. Bilgi of Istanbul University. Ikiztepe (dated from 4250-3200 B.C.) yielded 4 artifacts which had between 0.69-2.16% arsenic, while the rest were unalloyed coppers. This changes dramatically at Ikiztepe mound I (2800-1900 B.C.) where the majority of copper-based artifacts are alloyed with arsenic ranging from 1-12%. This site is only one of a number in northern Turkey where analyses (by Özbal) of objects suggest that high arsenical copper was intended. It is apparent that consistently high levels of arsenic were functionally correlated with specific objects at Ikiztepe as well: Pins and needles contain at least average arsenic (3.14%), jewelry contains the highest arsenic at 12.6%, spearheads average 5.5% As with some as high as 12.2%, and three ornamental spearheads with relief decorations contain consistently high levels of arsenic (9.2-10.2%). Microscopic analyses of polished cross sections of the Ikiztepe arsenic-rich tools revealed an inverse segregation of arsenic, which results in a silver-colored artifact. Another object from Early Bronze I levels of Ikiztepe, a leaded copper ornament (72.6% Cu, 1.8% Pb), produced surprisingly high nickel (22.7%) (Bilgi 1984, 1990).

Bronze objects are recovered at three main categories of sites—in hoards, in graves, and in settlement sites. The quantity of bronze known from this period together with evidence for expanded mining of copper (Kaptan 1986, Giles and Kuijpers 1974) and silver ores (Yener *et al.* 1991, Wagner *et al.* 1986) indicate that the scale of extractive processes increased greatly. Many new categories of bronze tools were developed suggesting a marked specialization of tool kits (cf. Amuq; Braidwood and Braidwood 1960), and other categories became more abundant. Agricultural tools, tools for wood and leather working, and tools for a range of other crafts make their appearance. Weapons were developed, including swords, spearheads, maceheads, and axes. Hammered sheets of bronze metal made the manufacture of weapons, jewelry, and ritual artifacts much simpler. Bronze artifacts from this period include several distinct categories, comprising both personal ornaments such as pins, bracelets, rings, necklaces, pendants, and appliqués, and tools such as axes, adzes, and sickles which were used to clear forests, build habitation and fences, and harvest grain.

Weapons constitute a third category, the archaeological evidence indicating that the most decorative swords and other weapons found in public buildings were perhaps stocked there for use by a limited number of elite individuals. These objects required much more specialized skill, as well as diverse raw materials, than most personal ornaments and tools. Indirect evidence of weapons appears in seal impressions dated to Gawra XIA which depict tools and weapons such as the bow, spear, and a trident-like object (Tobler 1950: Pl. 163, 83, Pl. 163, 89). Two-part molds for shaft-hole axes are found at Gavur Höyük near Pulur (Koşay 1976) and actual axes were found at Karaz (Koşay and Turfan 1959). The oldest multifaceted molds were found in Arslantepe VI, Late Chalcolithic levels, together with shaft-hole axes, which suggest a two-piece mold. Bronze weapons became standard symbols of elite status and characterized rich finds throughout the Chalcolithic period in many regions.

A large proportion of the bronze objects from the fourth millennium B.C. come from hoards—deposits of bronze metal intentionally buried in the ground. For example, 416 objects were found in a cave at Nahal Mishmar in Israel dated to the Chalcolithic period, c. 3750-3500 B.C. (Bar Adon 1980). The development of techniques of working sheet bronze and gold deserves special attention. This new technology emerged during the Chalcolithic period together with all the other cultural changes of the period. Polychromatic effects, too, required special skill as well as effort, and the products of these techniques were intended for high-status individuals. Especially significant in powerful iconic imagery are metal male and female figurines bedecked in gold, silver, and electrum. Six made of tin bronze were found at Tell al-Judaidah in the Amuq, southcentral Turkey in phase G, which dates to 3000 B.C. (Braidwood and Braidwood 1960, Yener *et al.* 1996). These examples typify the role metal played in symbolic and religious expression and are objects of prestige, empowered with ritual significance.

During the fourth millennium B.C. metal was also accumulated as wealth once it was produced on a large enough scale. Metal objects could be easily stored and transformed into tools, weapons, decorative objects (bronze), and jewelry (gold). Unlike other types of wealth that may have been important earlier, such as land, livestock, or surplus agriculture, metal could be transported from place to place and conveniently secreted for safekeeping. "Luxury trade was not merely as a stimulus to production or an adjunct to stratification but also as a series of long distance exchanges of relevance to the capture of energy. Gold and silver were readily convertible into energy resources across much of the old world and their movement constituted a disguised transfer of essential goods" (Schneider 1977). The accumulated precious metals were easily transformable, liquid capital that could be used as payment in exchange for goods or labor (Wells 1984: 79-89). Bronzes found in Mesopotamia and Syria during this period attest to exchange networks, whether the sources are Iran, Turkey, or other areas (Stech and Pigott 1986). Tin, silver, gold, and copper occur only in certain parts of the Near East (see next chapter), and the wide distribution of metal objects in excavations attests to a metal priority in this trade.

Archaeologically, the most important transformation in metal production and exchange during the fourth millennium B.C. was this scalar increase. The change is most apparent in the huge quantities of bronze objects from the period that have been recovered and in the wide diversity of objects manufactured from the metal. A new wealth in metals seems to have been broadly disseminated. Bronze tools and ornaments were widely distributed throughout Anatolia and bronze was produced on a large scale and traded. The fact that many typical, modest settlements yield molds and bronze scrap that demonstrate on-site casting indicates that wealth was widespread enough to permit most communities to acquire bronze and produce their own ornaments and tools. The second case study demonstrates this threshold as well as changes in metallurgy accompanied by possible migrating populations.

Case Study Number 2: Arslantepe, Malatya

The multiperiod mound of Arslantepe is located within the city limits of Malatya in eastern Turkey. Measuring 200 x 126 x 26 m high, it is one of the largest sites (4 ha) in the Malatya plain. Excavations began in 1932-33, and continued in 1938-39 under the direction of L. Delaporte. C. Schaeffer excavated the earlier levels during the 1946-48 seasons. A new round of investigations began in 1961 with the Italian Archaeological Expedition headed by S. Puglisi, A. Palmieri, and most recently M. Frangipane.

The Late Chalcolithic level VII (radiocarbon dates calibrated 3700-3400 B.C., Alessio *et al.* 1983, Amuq F) has been excavated in the northeastern part of the mound. The more extensive and subsequent Late Uruk level VIA and Early Bronze I level VIB exposures are found in the southwestern area with numerous building levels. Dating for these levels is based on an internally consistent stratigraphic sequence and is supported by a lengthy series of radiocarbon dates (Palmieri 1981: 102). The marker for Syro-Mesopotamian contact, the beveled-rim bowl, first appears in a phase between Late Chalcolithic level VII and in the Early Bronze level VIA (calibrated radiocarbon dates 3700-3400 B.C.) and may indicate that there are intermediate levels. The early metallurgical industries are represented mainly in these levels. Seven period VII house building levels reveal a mudbrick niched architecture and wall decorations with a social structure in the process of becoming complex (Frangipane 1993a: 135). The artifacts, especially the metals, hint at a developing craft specialization. Mass-produced chaff-faced ceramics, the slow wheel, and potters' marks speak of a trend towards different organizational principles.

A new area recently excavated on the western slope (E6) revealed buildings directly below period VIA. These substantially large buildings occupied the then summit of the settlement and were monumental in nature. They had stone foundations and a white-plastered mudbrick superstructure. A stratified sequence of red and black wall paintings in a series of superimposed rooms showed functional continuity for the rooms from level VII to VIA (Palmieri 1978: Fig. 5). The ceramics, however, were dissimilar, showing a dramatic change from level VII chaff-faced and red burnished wares to level VIA Uruk-like and black-polished wares. Another terraced building (A563) with walls reaching a thickness of 1.2 m also had evidence of red- and black-painted geometric designs. In Room A617 copper fragments, pins, and small chisels as well as copper ore were found (Frangipane 1993a: 147).

In the following period, level VIA, Arslantepe is a settlement on its way to becoming a complex urban society with strong Syro-Mesopotamian influence. It is characterized by a surprisingly advanced metallurgical technology which appears in tandem with non-local and local management devices such as sealings, indicating a control over the production and/or storage of metal materials. The ceramic assemblage also reflects both an intrusive Uruk-inspired, wheel-made, often reserve-slip light-colored pottery, and a handmade red and black burnished ware typical of central and northeastern Anatolia. The Late Uruk ceramics occur within a wide context of local styles of pottery and sealing traditions; local cultural traditions co-exist with Uruk culture, a phenomenon apparent also in the upper Euphrates at varying times and sites (Frangipane 1993b; Frangipane *et al.* 1983). It is worth reiterating that a variety of local situations and external relations mediate the degree to which metallurgical production and innovative advances are part of the assemblages in the Anatolian Euphrates basin. The increasing markets spur a complex exchange in manufactured products of high craftsmanship.

Four large public buildings, I-IV, and partial remains of Building XVI were excavated on the upper slope (Frangipane 1992) in an area of 860 m^2. Built with stone foundations, buttresses, and a mudbrick superstructure, all were covered in mud plaster and often painted white. All bore traces of fire. Building I, called a temple, has a large, rectangular cella and two adjoining rooms on its north side. The cella has a podium in the center and a basin on a low platform in the back wall between two niches. Traces on the walls of painted decorations as well as concentric oval seal impressions (Frangipane 1992: Fig. 20) are reminiscent of Uruk decorative devices in southern Mesopotamia. A barrel-shaped lead bead was found in Room A77. The oldest building is Building IV, which may have been part of a large palatial building, with two principal structural phases. A silver ring with overlapping ends was found in Room A206 (Frangipane 1992: Fig. 63: 6 sample no. 128). A monumental gate (A181) has an entrance room with an elongated rectangular plan. A number of small copper objects were found by this gate, Room A206, and the adjoining corridor.

Another room, A364, had paintings of two stylized human and vegetal motifs on the wall (Frangipane 1992: Pls. 10, 11).

Building III, Room A 113 yielded 22 metal objects in the mudbrick wall collapse level (Palmieri 1981: 107). These were fabricated with high arsenical copper, and consisted of nine sword-like blades, three of them decorated with silver inlay, twelve spearheads, and a quadruple spiral plaque or ingot. They were found in two bundles and may have been hung on the wall (Frangipane and Palmieri 1983a: 314, Fig. 18). The spearheads belong to a group with a leaf-shaped blade, cylindroid mid-rib, long ovoid butt end, and a straight chisel-ended square tang, a group which has a large geographical distribution (Stronach 1957: figs. 8, 4); some of the wooden shafts were still preserved. The swords have long straight-edged blades, which have flat sections or truncated ridges and have a hilt with a semi-circular head. They are the earliest known swords and typologically no known parallel exists for them at this early date. The quadruple spiral plaque has a square cross section with ends split and spirals inwards. Found in the same room is the shaft of a silver pin which reiterates the increasing use of silver during this period.

The presence of the spearhead and sword at Arslantepe may indicate the use of metallurgical technology in war as is the case in Mesopotamia (Frangipane 1985: 220). The spearheads have a number of affinities with later leaf-shaped blades which have been found in Early Dynastic Susa and Tello. Although a spearhead from an Ubaid III grave at Ur (Woolley 1956: Pl. 30) is earlier than the Arslantepe example, two copper and silver examples from Uruk are contemporary. The Arslantepe spearheads combine two different characteristics also apparent in Mesopotamian examples: the shafting system through a chisel-ended square tang with a long butt and the leaf-shaped blade.[5] The quadruple spiral form has widespread popularity in the Early Bronze Age, especially in jewelry.[6] Double and quadruple spiral beads and pins of gold, silver, and copper have been found in a number of

---

[5] This type appears earliest at Arslantepe, and later in Early Bronze Age northern Anatolia at Dündartepe (Stronach 1957: 115, Fig. 9: 4), Ikiztepe (Bilgi 1984: Figs. 12, 33-36), and Horoztepe (Özgüç and Akok 1957: Fig. 13). This type also appears on the southern coast of Turkey at Cilicia Soli Pompeiopolis (Bittel 1940), Tarsus (Goldman 1956: Fig. 14), and Silifke (Bittel 1955, Fig. 10). EB III Carchemish (Woolley et al. 1952: Pl. 61) and Kara Hassan (Woolley 1914: Pl. XIXc, 2) in the Syro-Anatolian Euphrates basin are comparable examples further to the east. Especially important are the contemporary Amuq G tin-bronze figurines from Tell al-Judaidah of warriors carrying a mace, a leaf-shaped blade, and a poker-butted spear (Braidwood and Braidwood 1960: Figs. 240-242). A full-length blade was also found in Amuq phase H levels (Braidwood and Braidwood 1960: Fig. 293: 4).

[6] The motif appears first on seals and sealings from the Late Uruk-Jemdet Nasr period in North Syria and Mesopotamia at Tell Brak (Mallowan 1947: Pl. 19: 15), Amuq G (Braidwood and Braidwood 1960: Fig. 253: 7), Jebel Aruda (van Driel 1983: 53), and the subsequent period at Arslantepe, VIB (Palmieri 1981: 110, Fig. 10: 2).

Early Bronze Age sites. The closest similarity in both form and relative size are examples from Ikiztepe which were buried by the waist of a skeleton (Bilgi 1984: Fig 18: nos. 272-276).

New cultural elements appear in period VIB, which have similarities with an east Anatolian-Transcaucasian, Kur-Araxes origin.[7] An influx of handmade red-black burnished pottery (Frangipane and Palmieri 1983b: 536-42), new building techniques and decorated, elaborate hearths appear in this EB I phase (3100-2900 B.C., Amuq G). Widespread sites in the Erzurum, Malatya, and Keban area, such as Pulur-Sakyol, Korucutepe, Norşuntepe, Karaz, Pulur (Elazığ), Güzelova, and Tepecik, all display these elements. Perhaps brought by a migrating population, elements of this culture appear in the Amuq and coastal Syria and into Palestine, where it is known as the Khirbet Kerak culture. In a later phase, period VIB2 at Arslantepe, ceramic connections with the south, specifically the northern Syria and upper Euphrates region, are resumed. Amuq G-like repertoires appear with wheel-made plain simple wares, reserved-slip jars, cylinder seal impressions on pottery, and rectangular mudbrick architecture. These resumed connections with the south are also found in the metals.[8]

Perhaps more importantly metallurgical activity was documented for period VIB2 by the abundant quantity of slag and ore found *in situ* piled up in one of the houses (Palmieri 1981: 118). This suggests that the minerals were stored as well as worked in these structures. Both copper and iron oxides were found in this level (Palmieri and Sertok 1994: Fig. 4). A paved courtyard surrounded by an array of rooms yielded a crucible and multifaceted sandstone molds (Palmieri 1973b: Figs. 18, 19). Crucibles (Palmieri 1973a: Fig. 45: 1-3) made of clay and multifaceted molds for casting chisels and flat axes attest to the magnitude of metallurgical production at the site. The beds for a variety of utensils are carved on each side of the mold. Multifaceted molds such as these have been found in Tarsus (Goldman 1956: Pl. 436: 2), Amuq phase J

---

[7] Variously called Karaz-Pulur, Kura-Araxes, Transcaucasian, and Khirbet-Kerak, this cultural horizon is a highly complex configuration of pastoral elements and sedentary populations and a matter of much discussion. Typified by a highly polished red-black burnished ware, often round architecture and distinct metallurgical traditions, the material remains occur from the late fifth into the 2nd millennium B.C. Its origins, too, are a matter of much dispute, but a northeastern Anatolian-Caucasian homeland is posited. On the basis of the appearance of these cultural elements earliest in the north, and progressively later in southeastern Turkey, and finally in Israel in EBIII it is possible that part of the population migrated from the north—*see* Sagona 1984 with references.

[8] Three pins from VIB with conical fluted and unfluted heads show stylistic parallels with Carchemish (Woolley *et al.* 1952: KCG 1 and 2), Amuq H (Braidwood and Braidwood 1960: Fig. 292: 14), Norşuntepe (H. Hauptmann 1972: Pl. 69: 6), and Hassek Höyük (Behm-Blancke 1984: Fig. 8). Other jewelry with shared attributes are silver spiral rings (Palmieri 1973a: Fig. 47: 3). A shaft-hole ax has parallels with a widely known Syro-Mesopotamian type (cf. Amuq J; Braidwood and Braidwood 1960: Fig. 351: 9).

(Braidwood and Braidwood 1960: Fig. 350: 1), and Troy (Schliemann 1881, Blegen 1950).

Development of Extractive Metallurgy at Arslantepe
The investigation of the extractive metallurgy of Arslantepe by a team of scientists was a multifaceted approach to the role metal played at the site. First, the analysis by SEM with a Link Energy Dispersion System and ICP of excavated metal artifacts, ore, slag, and crucibles was done at Technologies Applied to the Cultural Heritage of the National Research Council, Rome by A.M. Palmieri, H. Hauptmann, and K. Sertok. Second, the local sources of copper in northeastern Turkey were investigated. Finally, smelting experiments at Arslantepe documented the processing parameters, such as temperature, flux, airflow, quantities and types of copper ore, and the resulting residues. A total of 85 samples of ores were analyzed from the prehistoric periods.

In the earliest Chalcolithic level VII artifacts, it is apparent that alloying has already been achieved. Five analyses of the chisels and awls from these levels show that two were pure copper, while three showed arsenical alloying. One arsenical copper chisel (Caneva and Palmieri 1983: Table 1, no. 353) contains 2.47% arsenic as well as appreciable levels of nickel (1.29%) and bismuth (0.81%). The 1% bismuth in sample no. 347 and the 1.14% iron and 0.68% nickel in sample no. 354 suggest the regionally characteristic experimentation with polymetallic ores, especially those containing arsenic. The preference for polymetallic ores is easily observable in the ores found in these levels and suggests that copper-enriched ore zones containing arsenic minerals such as fahlerz were being extracted. A chrysocolla ore contained As (7.52%), Sb (7.13%), Fe (5.53%), Cu (25.6%), and traces of zinc, bismuth, and nickel as impurities (Palmieri, Sertok, and Chernykh 1993: no. 306). While another ore had similar As and Sb contents, and contained Ni (1%), Pb (53.83%), and Cu (7.8%).

In the subsequent level VIA, the crafting of blades and spearheads reveals a number of sophisticated metallurgical techniques. According to metallographic analysis, blades were cast in open molds and spearheads in closed ones. In some of the swords (Caneva and Palmieri 1983: 649: Table 1, no. 12), arsenic contents vary from one side of the blade to the other. This is indicative of selective dispersion effects where the metal in contact with the mold cools more slowly than the one facing the air and thus contains more arsenic. Minute differences in the details indicate that they were not cast in the same mold. After casting, the edges were hardened by cold working and sharpened by hammering and annealing which resulted in harder swords than spearheads (Caneva, Frangipane, and Palmieri 1985: 115-20). The crescentic-shaped edge of the hilt, where it joins the blade, was decorated with horizontal bands of triangular and zig-zag patterns. The triangular spaces were chiseled and inlaid with silver. The swords range from 46-62 cm in length and from 410-960 g in weight. All the hilts have curiously flat sections making them difficult to grasp, suggesting that they may have been ceremonial. However, the careful edge sharpening certainly suggests use as a weapon. The spearheads are less standardized and range in size from 41 to 53 cm in length.

The analysis of the level VIA artifacts demonstrates that weapons were made of arsenical bronze with arsenic contents ranging from 2-6.5% (Caneva and Palmieri 1983: Table 1). This introduces and echoes the subsequent Early Bronze Age technological tradition of high arsenical bronzes shared by the Black Sea metalliferous mountain regions in the earliest stage of alloying metallurgy. Consistent percentages of arsenic are functionally specific in the swords (from 3.2-5.8%), spearheads (1.3-4.3%), and the plaque (5.6%). A bimodal distribution is indicated for the differences between arsenic contents of swords versus spears (Caneva, Frangipane, and Palmieri 1985: 117) (Fig. 4a); thus two separate stages of smelting may be indicated. However, the imprecision of metal-working techniques at this time may make these distinctions irrelevant (Caneva and Palmieri 1983: 639).

It appears that either arsenic was intentionally added or arsenic-rich ores were intentionally used. A copper ore containing As (1.32%), Ni (3.58%), and Fe (3.32%) was found in this level. Arsenic levels in the small finds measured up to 8.23% in a fragment, 9.57% in the hilt rim of the sword, and 7.4% in the plaque rim, and suggest a choice of arsenic-rich ores for the silvering effect achieved by arsenical segregation. The weapons stand out as having high arsenic and no nickel (Caneva and Palmieri 1983: 641), a compositional difference which may indicate the intentional addition of arsenic ore. Chisels and awls ranged from 1-3% arsenic content, while several fragments showed no arsenic. Utilization of copper ores with some impurities or remelting may be indicated here. A ternary diagram of the trace elements in the artifacts suggests that most were derived to a lesser extent from oxides than from sulfides (Caneva and Palmieri 1983: 643) (Fig. 4b), although it is still difficult to be sure of the original ore used. The possible use of sulfides may indicate several stages of smelting although recent experiments have demonstrated that co-smelting in a crucible is actually possible and probably more advantageous (Rostocker, Pigott, and Dvorak 1989, Rostocker and Dvorak 1991). Given some of the complex ores that were found even in the earliest levels, the low iron levels seen in the artifacts suggest that the ancient smiths were very sophisticated in their smelting to be able to get rid of all the iron in the slag. Another explanation for the low frequencies of iron is that only oxides and carbonates are used (Caneva and Giardino 1996: 452-3), however, this does not explain the presence of iron-rich ores in the same levels. High nickel and arsenic values are also indicated in a possible nickel-arsenic sulfide ore as well as in the analysis of some slag samples (Palmieri *et al.* 1997: 61).

Both silver and lead artifacts are attested in period VIA including a relatively pure silver pin devoid of lead which may have been made from a silver ore (Caneva and Palmieri 1983: 650: Table 1: no. 33). One silver ring contains rather high levels of lead (2.76%) indicating that it was fabricated from an argentiferous galena or cerussite ore through a two-step cuppelation process (Caneva and Palmieri 1983: 650: no. 129). The 3.72-9.35% copper content is a typical addition to harden the silver. One arsenical copper pendant from VIA had a high lead (9.77%) content (Palmieri, Sertok, and Chernykh 1993a: 395). Tin also appears as a relatively high trace element in a lead bead (sample no. 143: Sn 0.42%, Zn 0.33%) and in many of the copper artifacts. The silver inlay from a sword shows up to 4.51% Bi and 1.01% tin (Caneva and Palmieri 1983: 649: Table 1: no. 30). These high tin values are not surprising given the tin-rich composition of the argentiferous lead-zinc-copper polymetallic Taurus ores such as those at Bolkardağ. The variety of ores found in the VII and VIA levels consist of galena and cerussite, lead ores, as well as complex copper sulfide ores such as tennantite, chalcocite, bornite, several with Fe-As-S phases, oxide ores, olivinite, Cu-Ca-arsenate, cuprite, malachite, and large quantities of iron oxides (Palmieri *et al.* 1996: 447). Use of polymetallic ores is also indicated in slag samples from level VII and use of a complex nickel ore with Ni-As-Sb.

In 1996, a tomb (c. 3000 B.C.) was found containing 75 objects made of metal, some stylistically paralleling the spears and blades from VIA. Splendid gold jewelry were analyzed to be silver and alloys of copper/silver. One dagger was fabricated from 50% copper and 50% silver, giving the object a silver-like appearance (Palmieri, Hauptmann, and Hess 1998).

Demonstrating the utility of technical analysis for pinpointing technological choice, analyses of polymetallic ores exploited in both levels VII and VIA (Palmieri and Sertok 1994: Figs. 6-13) show abrupt change with the advent of new cultural elements (Fig. 5). Significant changes both in the magnitude of production organization and style of technological choices appear in level VIB2 and are synonymous with transformations in architecture and other aspects of culture. Analyses of the metallurgical debris from these levels indicate that copper or iron minerals, or ores with an admixture of copper and iron, are now in great abundance (Palmieri and Sertok 1994: Fig. 5). The ores found in these levels are pure sulfides and oxides of copper and iron such as chalcopyrite/pyrite and cuprite/malachite/iron oxide/jarosites. Slag also contains Cu/Fe matte, rich in copper, some containing copper metal prills. These pyrites are very low in arsenic, antimony, and nickel (Palmieri *et al.* 1996: 448, 449, Fig. 1). Ten analyses of ores found *in situ* show copper ore with about 1.4% iron content and iron ore with 1.5% copper content. Ores containing about 1% Mg, Ca, and Na content also appear sporadically. There is a marked selection of copper or iron minerals or mixed copper and iron minerals (Palmieri and Sertok 1994: 123). A chalcopyrite ore with no arsenic was found in these levels (Caneva, Frangipane, and Palmieri 1985: Table 2). In fact, arsenic-rich, antimony-rich, or nickel-rich ores do not show up in these levels at all, despite their abundance in earlier and later periods (Palmieri and Sertok 1994: Figs. 7-9). Metal prills found in three crucible fragments from level VIB2 were analyzed and contained only Cu (20-36 %) and Fe (5%).

Interestingly the artifacts from level VIB do not show a change in the choice of arsenic alloying despite the change of ores used. Again, a few artifacts are pure copper and arsenic is still used from 1-6% in chisels and pins. Iron levels are low, with the exception of an obviously oxidized chisel which contains 1.64% Cu, and nickel is within the same parameters as copper (1-2.7%). The ternary diagram (Caneva and Palmieri 1983: 643) (Fig. 4b) does note a possibly greater diversity of ore types used on the basis of the trace elements of the artifacts. Oxides, relatively purer copper ores, were chosen, however, sulfides also seem to be used. Practically no lead is seen in minerals found in these levels, even though earlier and later periods show several percent lead content (Palmieri and Sertok 1994: Fig. 6). The discrepancy between the types of ores preferred and the alloys achieved could be resolved by either mixing an alloy rich in nickel and arsenic or deliberately using arsenic ores.

In the following VIC and VC periods (Early Bronze II and Middle Bronze Age), complex copper ores with arsenic and nickel originally exploited during the Chalcolithic are again found intra-site. An arsenopyrite ore from level VID with bismuth, antimony, and iron shows unusually high levels of tin (0.31%). It is obvious that arsenic ores were intentionally added to copper by this time as is evident in the selection of ores brought to the site (Palmieri and Sertok 1994: Fig. 13). An ore found in level VI in Room A604 contains As (8.81%), Sb (2.58%), Pb (33.28%) and Cu (14.94%).

In 1995, the metallurgical campaign focused on the Keban mines, and Zeytindağ in particular (Palmieri *et al.* 1996). The shaft and gallery systems here bear resemblance to the karstic, limestone cavities and infillings of Kestel mine in the Taurus Mountains (see below). That is, the ore (oxides and, to a lesser extent, sulfides) is easily accessible because the natural cavities provide simple extraction, indicating that these ores were possibly utilized with simple groundstone tools. Analyses of 23 ore samples yielded iron zinc, arsenic, with arsenic at 10%; gold (200 ppm) and lead was high as well.

Changes in metallurgical technologies at Arslantepe in the VIB period have been interpreted by the excavators as an exchange of copper oxides from the original homeland of incoming populations. In this view once connections were severed and the population changed, then the less desirable local ores were used again. However, an alternative scenario would be that technological styles predicated the changes in ore selection. The availability of complex polymetallic ores and the ability of the craftsmen to smelt it could have remained the same throughout the periods in antiquity. The choice was whether to use it or not since complex polymetallic ores were used at Arslantepe by both earlier Chalcolithic and later Early Bronze-Middle Bronze metallurgists. The change to the use of simpler, purer copper oxide ores reflects the technological choice of the level VIB metallurgists who must have used arsenic-rich ores for alloying directly.

The main characteristic of the metallurgical tradition entailed production of arsenical bronze and pure copper throughout the Chalcolithic and Early Bronze Age, however, alloying preferences varied in different periods. Complex arsenic-lead-antimony ores, such as sulfides and iron-rich polymetallic ores, were widely used throughout the 4th millennium B.C., in levels VII and VIA, while pure copper and iron ore were used as well in smaller quantities. Sulfuric fumes, slag heaps, and a much more messy operation necessitated that the ores be worked outside of the settlement proper. Thus in level VII and VIA very little slag or ore is found internally. The technology was so advanced that multistage processes for producing lead and other complex alloys had been perfected and the results are found at the site. The next period, VIB, shows a dramatic shift occuring in technology as oxides of copper were easily smelted in a crucible, leaving little slag.

Using these production parameters derived from analyses, several smelting experiments approximating archaeological precedents for furnace and crucible use were performed (Caneva, Palmieri, and Sertok 1989, 1990, Caneva, Sertok, and Palmieri 1991). Earlier attempts with stone and mud furnaces, bellows, and sulfide ores were not as successful as crucible smelting over an open fire surrounded by stones (Palmieri, Sertok, and Chernykh 1993). The ore used was a low-grade copper oxide/sulfide ore from Çayırköy. Successful smelts were achieved with 5 kg of charcoal, preheating the crucible, 1-3 mm grain size charge, and the use of handbellows for 20-30 minutes. Slag cakes 6-9 cm in diameter were produced weighing 200-300 g. Prills of copper-sulfide matte and metallic copper were formed inside the slag, some having dropped through the charcoal bed to the base of the crucible. Laboratory experiments to make arsenical copper alloys were also attempted. This was achieved using realgar (an arsenic mineral) which lowers the copper smelting point to 830° C and produces up to a 9% arsenic alloy (Palmieri, Sertok, and Chernykh 1993). A similar slag cake from Period VII was found and analyzed at the Institute für Archaometallurgie at the Deutsches Bergbau-Museum at Bochum. A thin section of the slag revealed that it contained primarily lead silicates (20-30% Pb), iron, and calcium. Prills consisting of copper-arsenic-antimony-lead-nickel alloy were found inside the slag cake. Although the type of metal produced is not yet certain, it is apparent that they were smelting polymetallic lead ores.

In sum, Frangipane (1998: 70) notes that while the site economy is primarily based on agriculture, both the organizational structures put into place as evidenced by seals and sealings and the quality and quantity of metals produced marked the fluorescence of this site.

The Altınova Valley Sites: Keban Dam Salvage Projects

The next section introduces a valley, the Altınova near Elazığ, which has yielded a number of sites most of which showed that intra-site metal working was one of the important activities during the Chalcolithic period. Norşuntepe, the largest site in the Altınova valley, is located 26 km southeast of Elazığ and provides representative evidence of metallurgical activities. The mound rises 30 m over the alluvial plain; the total including the lower terraces measures 600 x 800 m and the summit alone is 140 x 110 m. The stratigraphic sequences provide detailed information from the Chalcolithic periods to the Iron Age. The site was excavated as part of the Keban Salvage Project for six seasons between 1968-1974 by H. Hauptmann under the auspices of the German Archaeological Institute. The third millennium levels at the summit were given extensive exposure (2700 m^2; K/L 19 levels 26-14), while a deep sounding (J/K 18/19 levels 10-1, measuring 20 x 10 m) on the west slope provided information about the earlier Chalcolithic levels. The ramparts of the EB I (level 16) defensive wall obliterated some of the relevant Late Chalcolithic houses on the slope, however, a number of multiroomed mudbrick structures were still extant (H. Hauptmann 1974: Pl. 72: 2).

Level 10 is the earliest Chalcolithic level and is dated by ceramics which include crudely made flint-scraped Coba bowls, pedestal bowls, and kalottenförmige vessels. Dark-faced burnished wares are present with one example indicating that graphite was applied on the surface (H. Hauptmann 1982: Pl. 36: no. 5). Ubaid-like painted pottery is also present (H. Hauptmann 1982: Pl. 36: nos. 7-11). The architecture of level 9 revealed a well-planned settlement with several mudbrick units (H. Hauptmann 1982: Pl. 35). Although the extent of the sounding (15 x 20 m) does not provide enough of the architectural plans, the units appear to be magazines flanking long rectangular central rooms, not unlike the tripartitite structures of Ubaid-related Değirmentepe. Again like Değirmentepe, the large central room yielded traces of black and red painting. Room 8 had a large furnace with a pit in front. To the west in Room 9 large quantities of copper ore, slag, groundstone tools, and animal bones were found. An open space to the east of Magazine Room 8 revealed a large amount of copper slag and alloying materials (H. Hauptmann 1982: 59-61, Pl. 20: no. 2). Three more furnaces (H. Hauptmann 1982: Pl. 20: no. 4) and quantities of slag were found in another room. Samples for analysis were largely taken from this level (see results below).

Subsequent Chalcolithic level 8 revealed a substantial mudbrick building with two niches cut into the wall and a large central hearth in the main chamber (H. Hauptmann 1974: Fig. 20). The walls were plastered white and a painting of geometric designs rendered with black and red pigments appears on the walls (H. Hauptmann 1976b). The level 7 niched building also has a painting of an animal made with black and red pigment (H. Hauptmann 1974: Fig. 21). Simple and fine-painted chaff-faced wares (H. Hauptmann 1974: Pl. 71) are similar to Amuq F examples and date this level to the Uruk horizon. Some signs of recording are evident in stamp seal impressions on chaff-faced wares with simple cross designs (H. Hauptmann 1974: Pl. 79: no. 2, 1976b: Pl. 50: nos. 1, 2, 3) like Değirmentepe. These resemble seals and sealings from Gawra XI-IX (Tobler 1950: Pl. 145: 385-388). Geometric designs appear on stamp seals made of frit (H. Hauptmann 1976b: Pl. 48, nos. 2, 4). Gable-shaped stamp seals with animal motifs were also found in these levels (H. Hauptmann 1976a: Figs. 42, 43, 1976b: Pl. 48: no. 1) with parallels to a number of examples mostly from the Amuq (Braidwood and Braidwood 1960: Fig. 191 no. 7). A bullae with a sealing of a horned figure (H. Hauptmann 1976b: Pl. 48: no. 3) again is indicative of storage and exchange.

In addition to the spirals, rings, awls, and hook found in Chalcolithic levels at Norşuntepe (H. Hauptmann 1976b, 1982), over 2 kg of copper ore and slag were found in a heap next to a smelting furnace/hearth in Room M (H. Hauptmann 1976b, Zwicker 1989: Fig. 22: 4A) dating to the Ubaid-related level 10 (H. Hauptmann 1982). Analyses of the metallurgical debris included X-ray, microprobe, and spectroscope, and were carried out at the University of Erlangen-Nürnberg, Germany (Zwicker 1977). Results indicate that copper production was extant using a polymetallic copper-antimony-arsenic oxide ore containing chalcopyrite (Zwicker 1991). A microprobe analysis of a Chalcolithic sample of slag revealed the presence of iron (magnetite), silicates, and copper in sandstone (Zwicker 1989: Fig. 22: 4B). The slag was rich in delafossit, magnetite, cuprite, and piroksin and poor in fayalite and wustite. The researchers point to the fact that the smelt was not achieved in a good reducing atmosphere which is typical of smelting in a crucible.

One sample of slag yielded arsenic contents that vary from 0.9% in the dark sulfide (matte) to 13.5% in the gray area (Zwicker 1991: 333, Fig. 6). They suggest that this is the product, speiss, of smelting an arsenide ore, fahlerz. Polymetallic ores such as fahlerz have a variable arsenic content throughout the ore body. Smelting it would also yield 0.5% to 2% Sb content (Zwicker 1989: 193). A number of suggestions have been made as to how arsenic was introduced into the alloy, thereby improving the castability of the copper (arsenical bronze defined by them as 0.5% or greater As content), and as to how arsenical bronzes were produced. Differing arsenic amounts can be introduced into copper from fluxes, ores, or slag during the smelting (Tylecote, Ghaznavi, and Boydell 1977). Native coppers containing arsenic, such as a native copper-arsenic ore from Talmessi, Anarak which melts at 1000° C and contains 3.7% As, would lower the melting temperature. An arsenical bronze experimentally made from this native metal contained between 2.5% and 21% As (Zwicker 1991: Fig. 3). Copper oxides (azurite and malachite) also contain arsenic. For example, ore from Laurion contains between 6% and 35% As (Zwicker 1991: Fig. 5). A smelting experiment was conducted with the reduction of copper under charcoal at 1250° C. The resulting metal contained 2.5% arsenic and this method is suggested for the Norşuntepe examples. Arsenical bronzes could also be made with arsenic ore co-smelted with malachite at 1150° C introducing arsenic into the blister copper. By heating in a crucible covered with charcoal the alloy is produced after a half hour. Other arsenical alloys were attempted including introducing nickel arsenides into liquid copper, which, with the addition of a CaO flux, yielded 1% As content (Zwicker 1991). Pure arsenic dissolves easily in a crucible with copper foil, covered with charcoal, and heated slowly to 800° C. A eutectic produced like this would melt at 689° C. A third method, using realgar (AsS), could be used to produce arsenical bronze. Tensile strength of castings with 5% ore containing realgar increased from 172 N/mm^2 to 238 N/mm^2. It is often difficult to ascertain which of these methods was used since not very many slag or crucible examples were found. In addition, very little slag is formed, often in the form of powder, in smelting oxide ores. Nevertheless, all the viable techniques are possible given the available technology at the site.

Lead isotope analyses were conducted on 3 ore samples, 5 slag samples, and one copper metal artifact from Rooms AB, M, Y, Ma, and V from level 10, dating to the Ubaid-related Chalcolithic period, by the Max Planck Institute in collaboration with Mainz (Seeliger *et al.* 1985: 641-2). Their analysis did not identify the source of the sandstone copper ore (no. Tü 36f). The Max Planck group, however, noted an isotopic similarity between this specimen and ore from both the copper source of Ergani Maden (Seeliger *et al.* 1985: their number TG 176C-1) and from Kısabekir near the Black Sea (Seeliger *et al.* 1985: their number TG 177A-1) located 50-60 km from Norşuntepe. In another study, statistical reassessments of these ratios and new analyses of Turkish ores by a Smithsonian Institution-National Institute of Standards group suggested that this ore also had probabilities (19.5%-46.6%) of belonging to an ore group from the central Taurus (Yener *et al.* 1991: 557: Taurus 2B group). The suggestion of Ergani Maden or Kısabekir may be correct but neither of these mining complexes has yet been characterized sufficiently well to allow probabilities of the sample relating to them to be calculated. Moreover, the Ergani Maden specimen which was analyzed nearly overlaps with the Taurus 2B group which would account for the similarity. The fact that the metal artifact and other slag samples from Norşuntepe (nos. Tü 39b and Tü 40g) are consistent with Ergani Maden ores (Seeliger *et al.* 1985: Fig. 30, Sayre *et al.* 1992: 104) suggests that the eastern sources were probably the source of Norşuntepe artifacts. Other slag samples plot in a number of ore groups from Küre and Tirebolu in the Black Sea (Seeliger *et al.* 1985: Fig. 30: Tü 38b, Tü 37b). Two of the ores (Tü 34a and Tü 34b) also come from an as-yet uncharacterized ore source. All of this demonstrates the multiplicity of sources tapped into by the smiths.

Sterile sand separates Early Bronze I levels from Late Chalcolithic levels. Small, single-roomed mudbrick houses supported by wooden posts and some wattle and daub constructions are reported (H. Hauptmann 1976b: Fig. 29, 1982: Pl. 17: no. 3). The structures at Norşuntepe exemplify the extensive use of wood, which is not surprising since analyses indicate extensive forests were prevalent in this region at this time. Round houses found in association with characteristic red-black burnished wares and distinctive andirons (Diamant and Rutter 1969) are characteristic of the "Early Transcaucasian" culture. Other wares include Amuq phase G-related pottery with reserved-slip decoration (H. Hauptmann 1972: Pl. 68: 1). Sixty percent of the ceramics are Uruk-related, including Syrian ware, that is, buff simple wheel-made pottery with reserved-slip surface treatment.

The metallurgical industry so prevalent in the earlier Chalcolithic levels continues in EB I houses. Crucibles appear from the earliest level 26 and subsequent levels 22-25 yielded a number of garbage pits full of casting ladles, crucibles, and copper slag. A domed, natural draft furnace for smelting was found in later level 21 in a substantial mudbrick structure.

The furnace which measured 60 cm in diameter was shaped like a keyhole with a small trough leading to a hollow pit full of ash (H. Hauptmann 1982: Pl. 18 no. 4, Pl. 31). Casting ladles, crucibles, and copper slag were found nearby on the street suggesting a metallurgical function for these rooms. In the later level 19, horseshoe-shaped ovens were found inside large posthole houses which may have functioned as metal workshops (H. Hauptmann 1982: Pl. 30). Next to a horseshoe shaped oven two-piece molds for making a shaft hole ax and five ceramic cylindrical cores for casting the shaft were found on the floor (H. Hauptmann 1982: Pl. 17: nos. 5, 6, Pl. 26: nos. 9, 10). Copper-based pins and a ring (H. Hauptmann 1972: 114, Pl. 69: 6) were the metal objects from these levels. Crucibles were found in great numbers in level 19. Later samples of slag, dated to 2800 B.C., were found to contain higher amounts of Co, Pb, Cl, and Zn, as well as sulfide which suggested that sulfide ores were being smelted, a change from earlier periods (Zwicker 1977). In sum, Norşuntepe was technologically capable of smelting polymetallic ores from its earliest levels (for detailed analyses of the slag *see* A. Hauptmann *et al.* 1993).

Equally impressive strides in polymetallic smelting metallurgy comes from the neighboring site of Tülintepe, located in the western part of the Altınova valley, in Elazığ (Arsebük 1983). Excavated by Istanbul University under the direction of U. Esin, it measures 300 x 200 x 10 m and periods represented at the site range from the Chalcolithic through the Islamic. Mention must be made of a hematite macehead found on the surface but stylistically dateable by type to Amuq phase D (Esin 1976b: Pl. 1c). Atomic absorption and wet chemical analyses were conducted on 105 objects, crucibles, ores, and slag from Tülintepe (Özbal 1983, Kunç and Çukur 1988b, Çukur and Kunç 1989). Analysis of slag from Chalcolithic levels (Amuq C/DE, Halaf through Ubaid periods) revealed high trace levels of zinc (Özbal 1983: 215: nos. Bü-26/82; Bü-30/82; 2.68% and 0.78% Zn, respectively). The high zinc nature of the slag continues into the subsequent Early Bronze I/II (Özbal 1983: 215: no. Bü-29/82; Zn 1.55%), suggesting that a source with zinc-rich deposits was exploited in both periods. The 4.02% and 22.11% copper content in the Chalcolithic slag samples suggest that the smelting process was neither standardized nor efficient. However, ten more slag samples and one lead ore (17.94% Pb) were also analyzed and these demonstrated that very efficient smelting was also attained with copper contents under 0.5% (Çukur and Kunç 1989). Iron contents range from 1.4-4% and are typical of crucible slags resulting from smelting oxidized copper ores such as malachite. This is further supported by the pyroxene ($CaFeSi_2O_6$), calcite, and quartz contents of the slag and the lack of fayalite (Bachmann 1982). A crucible with slag accretion was found next to a round furnace from Amuq F (Uruk-related) late 4th millennium B.C. levels (Esin and Arsebük 1974: 154, Esin 1976b: Pl. 1b). Ore with 45.59% copper content was also found (Kunç and Çukur 1988b: Table 2: no. 8) along with a number of slag samples that contained between 2 and 8.3% arsenic, attesting to the possible utilization of high arsenical copper ores (Çukur and Kunç 1989: nos. 1-5, Kunç and Çukur 1988b: Table 1: no. 16) and alloying with arsenic minerals. Kunç and Çukur (1988a: 100), on the basis of nickel content, suggest Ergani Maden as a source for the copper ores. Both Tepecik and Tülintepe iron-rich matte-slag samples indicate that either a chalcopyrite or an iron-rich flux was used (Özbal 1983: 215: Table 3: nos. 1-4, 5-9, 11-12, Esin 1984: 82).

Again within the fertile Altınova valley, Esin and her colleagues excavated the Keban dam salvage site of Tepecik, with deep sounding 8-O defining the Chalcolithic sequence. Ubaid-related painted pottery, which was made with the slow wheel, and local chaff-faced ware were found in Strata 24-18 (Esin 1972: Pl. 114: no. 2, 1982c: 14). Late Uruk-related architecture was found in the southwestern quadrant of the site, in squares 14-17 AB-A, 15-16/B, Buildings 1-2. A symmetrical tripartite building was revealed, with stone foundations which had been modified a number of times (Esin 1982: Pl. 69). The tell-tale Syro-Mesopotamia-related tripartite plan of the building escapes notice since a path, CF, was cut through Rooms FH, FG, FD, BL, BM, FI, and CL, presumably in a later period. Stylistically intrusive, Uruk-related beveled-rim bowls, reserved-slip ware, and plain simple ware are in copious evidence and occur together with local fruit stands, similar to those found in Alişar Chalcolithic levels and red-black burnished Transcaucasian ware (Esin 1982a: Pls. 72-4).[9]

Clumps of lead, copper, and slag and a clay crucible were found in levels 22 and 18, contemporary with late Ubaid/early Uruk (dated to Amuq phases E and F) (Esin 1976c, 1984, 1987). The crucible contained fragments of metal on its inner surface (Esin 1972: 157). Analysis of the iron-rich matte slags indicates that either a chalcopyrite was used, or that an iron-rich flux was added to the smelt (Özbal 1983: 215: Table 3: nos. 5-9, Esin 1981a). Tepecik yielded a number of iron-ore and slag fragments with high levels of copper (3-11% Cu) and a number of copper objects with high levels of iron (1% Fe), suggesting the use of polymetallic or sulfide ore sources. Very little copper had gotten trapped in the slag and there was a high presence of zinc (range 0.01-11.73%, Özbal 1983: 215). Good arsenical bronze was observed in another example (2.33% As) from Late Chalcolithic contexts. In addition, a sample of argentiferous galena (Özbal 1983: 214: no. Bü-34/82) was found at the deep sounding dated to late

---

[9] Typological analyses of the ceramics suggest that of the 91 Uruk-related vessel types found in the building complex, 41 types had parallels in Uruk, 24 were related to the Amuq, 18 to Susa, 16 to Tarsus, 11 to Hama, and 7 to Godin Tepe.

Halaf-Ubaid on the basis of ceramic similarities to Amuq phases D-E. Several awls, needles, and a double spiral pin were found together with slag fragments near a hearth in a building with Uruk-related pottery at Tepecik. The copper remaining in one Early Bronze slag sample was 2.62%, a less successful smelt. High nickel and arsenic contents (2.68% Ni, 4.82% As; Özbal 1983: 216: no. Bü-33/82) were found in an ingot of arsenical copper. Alloying with metals other than the usual tin and arsenic is evidenced at Tepecik where a pinhead was analyzed as having 1.8% antimony. Temperatures in the smelting process reached 1200° C. As shown at Tepecik, copper smelting could sometimes be very efficient in the Chalcolithic levels, although the standards varied widely.

A number of other sites in the Tigris-Euphrates basins revealed impressive metal finds although enumerating all the assemblages here would be unnecessarily long.[10] One worth mentioning in greater detail since analyses are quite extensive is Hassek Höyük. In the Ataturk dam area, Hassek Höyük yielded a number of metal objects in the Uruk-related Chalcolithic period levels such as copper pins (Behm-Blancke 1981: Pl. 13: h). Early Bronze I levels 4-1 contained bronze pins (Behm-Blancke 1981: Pl. 13, 1-3, 5), bronze weapons (Behm-Blancke 1981: Fig. 13, 5), and a pithos cemetary where 50 bronze objects dated to the Early Bronze Age were unearthed. In total 75 objects were analyzed by Max Planck Institute in Germany. The Uruk period examples were bronze pins with hemispheric heads (Behm-Blancke 1981: pl. 13: 1h, Schmitt-Strecker, Begemann, and Pernicka 1992: nos. HDM 1148, 1150, 1167, HASS 22) and these were analyzed for composition and lead isotope ratios for provenancing the lead content. They respectively contained 1.15%, 1.43%, and 0.87% As; one, HDM 1148, had 1.07% Ni content. These arsenic-nickel rich metals parallel the type of early low bronzes that were in existence at this time.

Numbers of metal objects were found in the EBI/II cemetery, situated 500 meters west of mound (Behm-Blancke 1984: Fig 9). Tomb gifts include a cylinder seal with a bronze zoomorphic attachment, stamp seals, spearheads, two flat celts, a chisel, a dagger, pins, a well-preserved macehead, a bronze bracelet, and a lead artifact. Typologically the spearheads are very similar to the ones found in Arslantepe VIA. The compositions of these did not show technological change from the earlier

---

[10] The site of Korucutepe yielded a number of late Chalcolithic objects including a hematite mace head, a copper blade, a silver armband with spiral ends, and silver spiral earrings from a grave (van Loon 1978: 399, Pl. 4: nos 1-2, 5, 400, Pl.: 4 no. 4), a silver bracelet with stamp seal of wild goat, and a diadem (van Loon 1978: 400, Pl. 5: no. 1, 3). One analysis exists from the important site of Samsat. A copper fragment from Chalcolithic levels contained 7.66% As, 1.7% Pb, 1.4% Sn, and 10.25% Zn (Çukur and Kunç 1989: Table 2, no. 1).

period. Eighty percent of the objects had between 0.5 and 5% arsenic content (Schmitt-Strecker, Begemann, Pernicka 1992: 110-111). The bronzes were also characterized by high nickel content, and are similar to Amuq F bronzes some of which have 10% nickel. Nickel concentrations of bronzes from Hassek, Norşuntepe, Mersin, and Tarsus dated to the Uruk and EB revealed that bronzes from Hassek have the most nickel with Mersin showing relatively comparable levels (Schmitt-Strecker, Begemann, and Pernicka 1992: Abb. 2). A positive correlation is demonstrated between As and Ni as well, suggesting that the arsenic and nickel were part of the copper ore. The sourcing information gleaned from the lead isotope ratios suggest Ergani Maden for the ores, although the nickel-rich metal source is still an open question.

The Mediterranean Coast

Metal and metallurgical advances are very visible in Anatolian sites along the Mediterranean coast as well. Yümüktepe/Mersin is located in the port city of Mersin and was excavated by Garstang from 1937-39 and 1946-47. Recent excavations began in 1993 under the joint collaboration of Veli Sevin of Istanbul University and Isabella Caneva of the Università di Roma. The site is large and imposing, 200 m in diameter and 25 m high. The new prehistoric excavations are concentrating on augmenting the information of the earlier periods from level XXXIII (Early Neolithic) to level XII B (latest Chalcolithic). This span has been dated by radiocarbon to a range from 7004-4046 B.C. (Caneva 1996: 6). Levels pertinent to metal finds include XVII-XII (Chalcolithic period, late 5th-4th millennium B.C.). Earlier occurrences of metal were mentioned above in a Neolithic context and small-scale ornaments continued to be made in the early 5th millennium B.C. An older painted-pottery tradition prevails in Mersin XXIV-XX; level XXII yielded a scroll-headed pin (no. 1703) and a nail-headed pin was found in level XXI (Garstang 1953: 76: Fig. 50).

Ceramics with Halaf decoration and a similar fabric were found in Mersin levels XIX-XVII where eighteen objects have been analyzed by Esin (Esin 1969: 145: nos. 17871-17888). Technological change becomes apparent with the first appearance of substantial tools which are not ornamental or decorative objects. There are striking instances of experimental alloying and low-level bronzes are evidently produced by mixing arsenic, tin, and, in one instance, lead. The level XVII foundation of Room 166 yielded a flat ax made of copper (Garstang 1953: Fig. 69 no. 1508) measuring 5.0 x 3.1. x 0.8 cm. The author notes that this is paralleled by objects which show clear traces of metal tooling and a chisel-marked stone (Garstang 1953: Fig. 67). Simple open molds probably began to be used at this time since flat axes and chisels begin to appear.

Also from this level, but from a less secure context, is a seal with a decorated base (Garstang 1953: Fig. 70). Garstang questions the attribution of this seal to this early date since it was found in the debris of this level and also because the seal resembles later examples. Intriguing is the fact that it contains 2.6% Sn, 1.55% Pb, and 1.2% As (Esin 1969: no. 17871), assuredly an experimental alloy. Geometrically decorated seal bases do occur in very early contexts in Anatolia, for example, in the Amuq (Braidwood and Braidwood 1960) and Koşk Höyük located immediately to the north of Mersin in Niğde (Silistreli 1990). However, the on-going excavations may yield additional and more securely dated examples of metal seals made with tin bronze from the Halaf period.

During the following level XVI, the site is fortified and then destroyed by burning. Fortifications include a circuit wall with stone foundations strengthened by offsets (Garstang 1953: Fig. 79) and an imposing gate with flanking extramural towers to guard the flank facing the river. The local painted pottery has Amuq E affinities and continues into XV-XIIB. Ubaid-related ceramics and large-scale architectural features appear in levels XVI-XIV (Amuq phase D, Halaf-Ubaid transition) and date to approximately 4909-4730 calibrated B.C. Substantial weapons and tools, as well as ore, appear in the structures of this level: a chisel, axes, an adz, ore, a polished tool, and six scroll-headed pins. The metal yielded evidence of intentional alloying with arsenic and evidence of the production of larger-scale artifacts. A transition is reached between the earlier manufacture of pins and luxury items and the later manufacture of heavier tools and weapons (Garstang 1953: 108, figs. 69-70, 109, 132: Fig. 80b: 137, 139: Fig. 85, 140, Esin 1969: nos. 17871, 17877, 176882, 17884, 17885, 17909). A much more precise sequence for the transition is now being worked out through the new excavations directed by Sevin and Caneva (Caneva 1996, 1998).

A large central building (Rooms 166, 175, 170, 180) lies to the south of the gate. Although the erosion of the site has obliterated the western edge of this structure, enough remains to suggest that this is a tripartite, Ubaid-style public structure, similar to ones found in Değirmentepe. Similar Ubaid features are four magazine rooms flanking a rectilinear central Room, 166, which contains a large hearth. Garstang (1953: 134) suggests that other magazine rooms would have flanked the western side, similar in plan to Gawra level XV. A number of the rooms of this large structure contained metal objects, again paralleling the Değirmentepe buildings. These include an ax head and a loop-headed pin from Room 169 (Garstang 1953: Fig. 80b no. 1323, Fig. 85 no. 1325, Esin 1969: no. 17875) and a polished metal tool or pin from area 177 (Garstang 1953: 140). Two pins were found in courtyard 189 (Garstang 1953: Fig. 85 no. 1331, 1330, Pl. XXI, Esin 1969: nos. 17877, 17878, respectively). Pin no. 1331 is important in that it is a low-level, perhaps experimental, alloy containing 0.75% Sn and 1.1% As. Courtyard 189 also yielded a chisel with a tapered tang that interestingly has a rivet hole (Garstang 1953: Fig. 80b no. 1329, Esin 1969: no. 17872). Room 184 along the fortification wall yielded a scroll-headed pin (Garstang 1953: Fig. 85 no. 1332, Esin 1969: no. 17879) and a copper adz (Garstang 1953: Fig. 80b no. 1334, Esin 1969: no. 17874). The dump yielded one more scroll-top pin (Garstang 1953: Fig. 85 no. 1333, Esin 1969: no. 17876). The larger artifacts were probably made with a simple open mold. It is possible that a workshop casting metal tools and weapons existed at the site. This is also suggested by the discovery of a fragment of copper ore in Room 179, unfortunately not analyzed.

Level XVb (Ubaid-Uruk-Amuq D), which also contained Ubaid-related ceramics, yielded a broad-ended chisel from Room 164b (Garstang 1953: 167 and Fig. 95B no. 1207, Esin 1969: no. 17880). Garstang discusses the less secure find place of this and the similar, but earlier chisel from level XVI, which came from a sealed context. Traces of hammering were evident and Garstang suggests that they were cast in a mold. Two long needles of copper were found in levels XIV-XIII (Garstang 1953: 167: nos. 1167, 1168, Fig. 108, Esin 1969: nos. 17886, 17885), but Garstang cautions that they may be intrusive. Bronze needle no. 1168 is interesting in that it contains 1.5% arsenic, making it an arsenical bronze alloy. Less securely provenanced are a bronze toggle pin with 1.3% Sn and 1.15% As from levels XIV-XIII (Garstang 1953: 167, unpublished no. 1313, Esin 1969: no. 17884) and a bronze awl attributed to level XIV (Garstang 1953: unpublished no. 1169, Esin 1969: no. 17882). Analysis revealed that these bronze alloys are ternary bronzes with 2.1% Sn and 1.25% As.

One of the interesting aspects of these early bronzes is the consistent use of both tin and arsenic in the same artifact. Perhaps these alloys are indicative of early experimental combinations of different ores. Equally consistent is the range (from 0.02-0.26%) of silver as a trace element in the copper objects. The discovery of tin traces in association with silver and lead gives rise to questions about the subtle relationships of the various other metals in a complex ore body and the technology of early bronze production. The Mersin early bronzes argue for an initial experimentation with polymetallic ores such as the ones from the Taurus, in other words, directly smelting the minerals in order to approximate natural alloys.[11]

---

[11] Other examples of late Chalcolithic metal finds close to the coast come from the Amuq site of Tell al-Judaidah (Mixed Range Amuq C-F). These included a reamer (0.9% Ni, 1.35% As), needles, chisels, pins, maces, and knives. Phase F yielded 7 reamers, 1 pin, 1 blade, 1 projectile point, and 2 chisels. The blade had a prominent midrib and four rivet holes (Braidwood and Braidwood 1960: 244-246, Fig. 185 no. 5). Semi-quantitative

C. The Early Bronze Age: Industrial Production

The Early Bronze Age, the 3rd millennium B.C., in Anatolia is characterized by dramatic political and economic changes on both regional and interregional scales. The changes are defined only archaeologically since writing has, as yet, not been found, despite hundreds of tablets from contemporary Syria and Mesopotamia. Changes are revealed in shifting population densities, the construction of monumental buildings on fortified citadels, migrations, a much more emphatic social stratification, and shifts in a diverse array of technologies.

In terms of metallurgical technologies, the mid-to-late-3rd millennium represented a renaissance of industrial metallurgy in the ancient Near East. This phenomenon has been characterized as a technical and industrial explosion (Ekholm and Friedman 1979: 47), during which arsenic and tin bronze became the major medium for fabricating complex artifacts (Watkins 1983). The relative frequency of copper goods in graves and the profligate use of other precious materials is impressive (Stronach 1957, Moorey 1982, 1994). In Mesopotamia, texts mention a diversity of metals and abound in formulas to fabricate them (Muhly 1973, 1976). Multitudes of objects demonstrate the use of hammering sheet metal, alloying, and casting of tools, weapons, ornaments, and statuary (Moorey 1994, M. Müller-Karpe 1993). The royal tombs at Ur (Woolley 1934) have yielded great quantities of gold, electrum, silver, bronze, and copper objects. Sheet metal was chased and repouséed, fittings were cast, riveting was employed, objects were soldered with tin (Craddock 1985), and filigree and granulation became commonplace throughout the Near East (Maxwell-Hyslop 1971). Burial and hoard assemblages from Troy (Schliemann 1881, Blegen 1950, Blegen et al. 1951), Alaca Höyük (Koşay 1944, 1951, Koşay and Akok 1966, 1973), and Ikiztepe (Bilgi 1984, 1990) reveal exquisite jewelry that only labor-intensive techniques could have possibly produced.

Casting with a lost-wax technique began on a small scale, with figurines, and increased in quality and quantity when larger statuary made its appearance. Two-part molds for shaft-hole axes were found in Norşuntepe (EB IIIa) and Gavur Höyük near Pulur. Actual axes were found at Karaz, Ahlatlibel, Alaca Höyük, and Kültepe. The oldest multifaceted molds were found in the Arslantepe VI Late Chalcolithic levels, together with shaft-hole axes, which suggest a two-piece mold. Multifaceted molds are also found at Troy II, Beycesultan IX, Alişar, Amuq phase J, and Tarsus EB II. Other examples on a smaller scale are two-part closed trinket molds (usually of stone) which were found in a number of contexts in

---

analyses of reamer 2 revealed 2.73% Ni and 2.05% As. In fact nickel was high in the pins, daggers, and chisels as well (Braidwood, Burke, and Nachtrieb 1951: 89).

Anatolia, southern Mesopotamia, and the Aegean and have provided a framework for postulating the mechanism for the interregional dispersal of the represented artifacts (Canby 1965). Suggestions for this mechanism include itinerant tinkerers who carried the trinket molds as part of their repertoire or the existence of an Anatolian clientele with interregional connections. The carved objects manifest a stylistic diversity and a mix of regional styles linking widespread geographical areas.

Polymetallism and polychromatic effects on artifacts became widespread in the 3rd millennium and were achieved by altering alloying materials, inlaying colorful stones such as lapis lazuli, carnelian, obsidian, and agates, or mixing a variety of metals together. Two of the effects are the shimmering silver quality of high-arsenical coppers and the red-to-gold color achieved by varying tin contents in bronzes. Analyses of objects from Troy demonstrates that gilding was also used in this period, while an artifact from Karataş demonstrates that silver casing was used to embellish an otherwise ordinary copper pin (Yener, Jett, and Adriaens 1995). The range of Early Bronze Age copper- and silver-working techniques (including lost-wax casting) reflects a long period of indigenous development and experimentation with a wide range of ore bodies. The establishment of a silver standard in Mesopotamia (Powell 1990) at this time ushers in the proliferation of silver and lead artifacts in the source areas.

The extraction of silver is achieved by the cupellation of galena (lead sulfide, PbS), often the primary ore utilized, although cerussite may also have been used in the earliest periods. A hoard of 16 silver ingots from Mahmatlar revealed high levels of zinc, which is also echoed in the contemporary Alişar "copper age" levels which yielded a lead pendant with 2.3% zinc, suggesting the smelting of polymetallic ores for silver. The refining of gold and silver (Prag 1978, Patterson 1971) and the cupellation of lead sulfides are in evidence in a number of sites. Silver ingots can be found in the Early Bronze Age at Troy II in Treasure A, which includes a great variety of silver objects. One unusual silver alloy ring from Troy contained 50.6% Ag, 1.4% Sn, 1.4% Fe, and 0.8% Cu. Jewelry hoards are characteristic of this period; a number of silver artifacts were found at Eskiyapar and great concentrations of silver abound in the "royal" graves of Alaca Höyük, at Horoztepe, and in the hoard at Mahmatlar. Analyses by Özbal of two of the Mahmatlar ingots (6.22% and 13.1% Zn) and Trojan silver jewelry (Zn ranges from 1.27-3.63%) and ingots yielded unusually high levels of zinc (Yener *et al.* 1991, Sayre *et al.* 1992). Two ingots from Troy were analyzed showing 4.5% and 0.9% Zn, and 1.8 and 1.1% Cu, respectively. This suggests the very early use of polymetallic silver ores. Silver-copper alloys were also utilized and good examples are found at the Alaca Hüyük royal tombs as well as the new burials at Arslantepe. One analysis of a silver cup showed 15% Cu content. In Troy one of the silver ingots contained 3.4% copper.

An analytical program, conducted by the Stuttgart laboratories with samples provided by Esin (1969) revealed the following information. Trace element analyses of copper-based assemblages from excavated contexts in Turkey indicate that of the 750 copper-based artifacts analyzed dating from this period, 600 had more than 1% deliberate additions of arsenic, lead, or tin. By the 3rd millennium B.C., 69% of copper-based artifacts had some form of tin or arsenic alloying. The increase in the amounts of silver in Anatolian contexts occurs synonymously with the use of high arsenical bronzes, which also gives the object a silvery color. High arsenical bronze alloys were a deliberate choice especially in the later period, the Early Bronze Age. Evidence of this technique was discovered when a bull figurine from northern Turkey in the Boston Museum, thought to be silver plated, was found to have a rich surface of arsenic (Whitmore and Young 1973). Arsenical copper use continued through the second millennium B.C. (Moorey 1994, Craddock 1985). Regional metal industries (Yakar 1984, 1985) took on a much greater role. Thus, if enough artifacts are analyzed to establish a metallurgical cross-section database, it may be possible to discern technological style zones. Mastery of the arts of smelting, melting, annealing, forging, working sheet metals, and alloying were all part of the metallurgical techniques perfected during this time (Maxwell-Hyslop 1971, Franklin *et al.* 1978).

The use of iron (Yakar 1984, 1985, de Jesus 1980, Wertime and Muhly 1980) took impressive strides during the third millennium B.C. Iron minerals were first used in pigments such as ochre (see Chap. 2). Chunks of hematite ore (iron oxide) were shaped into maceheads and hammerstones, such as the ones found at Korucutepe, Göltepe, and Tülintepe in the third millennium and even earlier in the Ubaid levels at Tell Kurdu (Yener *et al.* in press). This is not surprising since many of the deposits in Turkey contain massive iron components. Metallurgically, however, the use of iron changed from the making of trinkets from meteoric iron to the crafting of large-scale terrestrial iron objects in the 3rd and 2nd millennia B.C. This knowledge was put to use in both decorative and utilitarian objects. Small quantities of iron objects from Bronze Age contexts come from Tarsus, Alişar, and Kusura (Waldbaum 1978, 1989). The most important objects are from Early Bronze Age Alaca Höyük Tomb C, where two iron daggers with gold handles were found. The microscopic distribution of carbon in these items suggests that they were forged. Analyses of two more Alaca iron objects, a crescent-shaped plaque and a pin with a gold-plated head, yielded 3.06% and 3.44% nickel content, respectively; these may have been made from meteoric iron, although there are some terrestrial iron ores which do contain nickel. Other iron artifacts from contemporary sites are an iron ring (72.8% Fe, 6.12% Cu) and an iron fragment (73.32 % Fe, 2.19% Cu) from Tepecik.

In summary, the metallurgical technologies of the 3rd millennium B.C. represented a production that was fully developed and multiscale. Metal-working workshops in the urban centers finished and refined the pre-processed rough smelts that were assuredly produced in the highlands close to the mines. This is documented by the establishment of special-function sites in the mining regions of Anatolia, where the first tier of mining and smelting occurred. The following section presents the results of surveys and excavations in two major metal production areas of the central Taurus mountains, Bolkardağ and Çamardı, dating to the Early Bronze Age.

CHAPTER THREE

KESTEL MINE AND GÖLTEPE

The Problem of Tin Sources

If there is a single concept that has most unsettled the commonly held view of technological advances in metallurgy, it is that tin, a vital component of the then "high-tech" industry of its age (bronze), has been found not in an exotic, elusive place, but in the middle of a region where tin bronzes appeared prominently in the late fourth millennium B.C. Prior to this, most theorists had concluded that Anatolian and all other Near Eastern tin bronzes were made with tin imported from elsewhere (even in the early stages) and had proposed elaborate long-distance exchange systems with presumed sources of supply. These sources were assumed to be in Malaysia or Cornwall (Muhly 1973: 262-88; 409-12) or in the Hindu Kush mountains of northern Afghanistan (Cleuziou and Berthoud 1982, Franklin *et al.* 1978).

In 1985 high trace levels of stannite, a complex tin ore (Cu_2FeSnS_4), were discovered in analyses of ores from Bolkardağ in the central Taurus mountains (Yener and Özbal 1987) (Plate 1). Immediately following this find in 1986, cassiterite (tin oxide) was identified in three streams near the town of Çamardı, Niğde, forty kilometers north of Bolkardağ. As part of a major 5-year research project investigating the sources of gold by the Turkish Geological Research and Survey Institute (M.T.A.), cassiterite was identified after panning 80 tons of alluvial stream sediments in the Niğde Massif mountains. The Early Bronze Age Kestel mining complex was discovered on the slope 200 meters above the highest tin-yielding stream, Kuruçay near Celaller village (Yener *et al.* 1989, Çağatay and Pehlivan 1988, Pehlivan and Alpan 1986). An Early Bronze Age mining village, Göltepe, was discovered on survey in 1988 at the summit of a hill facing the entrance of Kestel mine. The galleries, quarries, and industrial processing/habitation sites were investigated by combined teams of geologists, minerologists, and archaeologists in the ensuing years, providing important information about a first-tier industrial production complex in the highlands.

Much heated discussion and passion has been unleashed by this recent finding of a major source of tin in Turkey. After the initial surprise, some in the scholarly community ignored the findings in the hope that they would go away. Others fearing the resulting paradigmatic shift displayed varying stages of dismay and disbelief. A cursory summary of the bibliography reflects the sustained scholarly dialogue, especially our articles with titles generally beginning with the words "Comments," "Reply to," or "Response to." Finally, as the technical discussions and instrumental analyses became increasingly more complex and no reconciliation of divergent views emerged, archaeologists awaited a final interpretive overview before integrating the impact of the findings into their reconstructions. Our discovery in the central Taurus mountains set the stage for unraveling one of the major unknowns which had long bewildered scholars working with metals in the Near East. It is important to point out that the Taurus sources are only two of probably many tin sources located in small, but significant, pockets in various areas of the Near East (Yener and Vandiver 1993a and b). A number of authors have noted the assays of other tin sources in Turkey (de Jesus 1980, Esin 1969, Kaptan 1983, 1995b), as well as possibilities of tin in the Caucasus (Selimkhanov 1978) and Yugoslavia (Taylor 1987).[1] Despite earlier dismissal (Muhly 1978), the tin mineralization in the Eastern Desert of Egypt has been taken seriously at last (Muhly 1993, Rapp *et al.* 1996). Good tin sources include Erzgebirge (Taylor 1983) and high trace levels occur in the ores from the Black Sea area (Tylecote 1981), Cyprus (Rapp 1982), and the Troad (Çağatay *et al.* 1982). These are fairly compelling indications that tin was more abundant in the Near East than was previously thought.

Aside from the disappointment of having a tin supply in a non-exotic location and the hint that multiple tin sources could have been exploited in the Near East, the following problems have been raised about the Taurus findings: 1) the relatively small amounts of measurable tin still extant in Kestel mine today; 2) the striking lack of tin slag deposits or furnaces at or near Kestel or Göltepe; 3) the seemingly amazing ability of ancient man to recognize the alloying material in the complex ore veins; 4) his equally striking ability to separate the tin from a low-grade, iron-rich tin ore and his even bothering to do so; 5) the lack of enough tin-bronze artifacts from fourth and early third millennium B.C. archaeological assemblages in Anatolia to account for this magnitude of tin production; and finally 6) the later Middle Bronze Age texts (19th-18th c. B.C.) which mention massive amounts of tin (11,000 lb.) brought into Anatolia by Assyrian merchants in the face of local tin sources.

A point reiterated by a number of critics is the low-grade level of tin extant in Kestel mine today and the fact that this was insufficient to be a major source for the Bronze Age (Muhly *et al.* 1991, Hall and Steadman 1991). The tin is indeed not obvious today since only sub-economic

---

[1] For Afghanistan tin sources see references in Pigott 1996; on new investigations in Central Asia see Alimov *et al.* 1998.

material remains unmined. Understandably, we are less confident in defining the original extent and tenor of the ore in an abandoned mine. To do so would be akin to quantifying the richness of the original gold veins of the 1849 Gold Rush by sampling the abandoned gold mines of California. According to geological and mineralogical reports, there have been two primary mineralizing episodes, an earlier tin-bearing and a later hematite one with weak tin (see below). The deposit was of considerable size and there was more than one period of mineralization; the most likely mineral mined both on the surface and underground was tin, but with the possibility of subsidiary gold. Today tin is present in approximately the 0.1-1% grade. Evidence of ore extraction continues below the marble into the underlying quartzitic schist and granitic pegmatites, with a total of 1.5 km of extraction tunnels explored to date. The underground galleries are extensive, measuring a minimum of 4600 cubic meters. Extrapolating from the low-grade ore composition with 1% tin content (what remains today for analysis), the space extracted would have yielded about 115 tons of tin.[2]

Puzzling, too, is how the prehistoric miners and those working the smelting could have recognized tin at such a low grade in the ore, assuming that the tenor of the ore was the same in antiquity. Admittedly, at the earlier stages of our own research, we had difficulty discerning the criteria for exactly how the tin was selected. It now seems likely that several methods were used. One was an age-old procedure, using a regular assay of samples—crushing, followed by panning as a guide to the ore. Cassiterite would separate out mechanically (see below). It did not escape us that gold may have been mined instead, since the stone tools used for concentration would be appropriate for tin or gold. The possibility must remain that gold was mined first or as well. If found in sufficient quantity, it would not have been neglected, although there has been negligible gold found at Göltepe, the processing site. In answer to the question of why bother to extract such a low-grade tin ore, the original ore would have been far richer than the powdered material found at the latest phase of Göltepe. Consequently, it is proposed that as the ore decreased in quality with extraction and demand for the new alloying material increased with the spread of tin bronzes, tin would have been as valuable as gold and certainly worth the effort.

Tin-bearing veins at Kestel are, in addition, easily distinguished in appearance from other veins and from the host limestone, in both color and texture. Color and texture are still important in the field identification of minerals and were a useful guide for the early miners as well. Especially distinct in appearance is the tin-rich hematite ore which has a gray-,

---

[2] Estimates which include the newly discovered galleries are described below.

sometimes burgundy, tinted, glittering appearance, unlike the much more matte appearance of hematite ore without tin. A large number of tin-impermeated hematite ore nodules were recovered during excavations at Göltepe and these resemble the ore from Kestel. Analyses of these nodules yielded an average tin content of 2080 ppm (with a range from 0-14,300 ppm), nearly three times the average still extant at Kestel mine. One sample contained 1.5% tin, suggesting that the tin originally mined at Kestel would have been at least a 2% or higher tin-rich ore, a very good grade by today's standards. This strongly suggests that only high tin-containing material was selectively transported from Kestel mine to Göltepe for processing (grinding) and smelting purposes. In order to recover the tin from the hematite matrix, the ore must have been crushed to a powdery consistency. The over 5,000 groundstone tools used in ore crushing from excavated contexts inside pithouse structures at Göltepe support this conclusion. Perhaps the best indication of processing aims is the undeniable increase of tin content from samples taken from veins in the mine, to samples from the hematite ore nodules found at Göltepe, and, finally, to samples of the multicolored ground and pulverized ore found stored in vessels and floors of pithouse structures. It is strikingly obvious that tin-rich hematite was being enriched on its path from the mine to the smelting crucible. None of the other elements analyzed showed this pattern of increase (Adriaens *et al.* 1999). But the answer to the puzzle of what was processed at these sites was finally unraveled when the production techniques were defined with the analyses of crucibles and ground ore powders, as well as with replication experiments (see Chapter 4).

As far as the question of the appearance of tin bronzes is concerned, there is no doubt that the early alloyed coppers found in Anatolia do contain tin. Whether intentionally added as metallic tin or as cassiterite mineral, tin was a part of the copper artifact composition; there is ample indication that some form of the element was involved in the production of metal objects. In the Amuq valley, the site of Tell al-Judaidah yielded unequivocal evidence of tin-bronze artifacts from the late 4th, early 3rd millennium B.C. (Braidwood and Braidwood 1960: 300-315, Braidwood, Burke, and Nachtrieb 1951). A pin and an awl from phase G contained 7.79% and 10% tin, respectively, and fragments of copper slag from crucible fragments had bronze prills with tin averaging 2 to 37% (Adriaens *et al. in press*). Six figurines made with tin bronze and festooned with multimetallic accoutrements point to a precocious ability to manipulate the local polymetallic sources. Other early examples of tin bronzes occur at Early Bronze I Kusura A, where several pins and needles show alloying with tin (Lamb 1936). At Kusura B levels, analyses indicated that 4 out of 18 artifacts sampled from the mid-3rd millennium levels have from 4.8 to

6.7% tin (Esin 1969: 136). Stòs-Gale, Gale, and Gilmore (1984: 26) have re-analyzed Anatolian Early Bronze Age tin bronzes sampled by Esin and have noted that these early analyses "underestimate the quantity of tin present by factors varying from 1.2 to 2.5." By the mid-3rd millennium B.C., relatively good tin bronzes are found in most areas of Anatolia and, perhaps even more relevant for Kestel, at sites along the Mediterranean coast. Located 80 kilometers south of Kestel, Tarsus Early Bronze II levels have revealed copper-based artifacts of which 24% are tin bronzes; in Tarsus Early Bronze III good tin bronzes are present as well (Esin 1969: 131-133). These bronzes contain up to 6% tin, and there are high-grade tin bronzes in the coeval phases H and I in the Amuq as well.

Kestel mine and the production/habitation site, Göltepe, went out of existence at the end of the third millennium B.C. One could speculate that local tin was no longer available in quantities sufficient to answer the increasing demands for this alloying material, especially as tin bronzes became more prominent. Purer, already packaged, readily available tin would have been attractive to metal producers who had long made tin bronze, although this competition would have devastated the local operations. Therefore during the Assyrian trading colony period (20-18th c. B.C.) it is not surprising that *annaku* (translated as tin—but see Powell 1990: 87) was being imported into certain Anatolian sites (Larsen 1976, Orlin 1970, Garelli 1963), despite the prior existence of local tin sources.

As more sources of metals are investigated, different production models, exchange patterns, and other socio-political and economic factors will emerge as effecting the circulation of these materials. For example, a restricted, more localized mining pattern typified by what geologists fondly refer to as "ma and pa" operations, exists even today in Turkey. The enterprising third millennium merchants could have been operating within a separate network, bringing in tin from an eastern source, perhaps Afghanistan, while Kestel or even other sources were supplying other regions.[3] It would not be surprising to find such a mosaic of interregional connections and commercial sophistication during this highly entrepreneurial period. It would be akin to the co-existence of a local and imported textile trade referred to in the Kültepe documents (Larsen 1976). Given the variable patterns of stability and political aggregation in Anatolia, northern Syria (Weiss 1986), and Mesopotamia during this time and the ample textual documentation of on-site metal technology and trade (Waetzoldt 1981, Waetzoldt and Hauptmann 1989), the central Taurus may have played a major role as a focus of competitive demands for metals. It

---

[3] This is certainly indicated by the third millennium B.C. textual documents from Ebla where tin was exchanged and was not expensive, suggesting alternative sources of tin (Archi 1993). The one-source-for-all model must indeed be re-examined.

is important not to lose perspective on intra-Anatolian commodity networks when postulating the appearance of exotic items from long distances.

Field Research in the Central Taurus Mountains: The Physical Setting

The central Taurus region was targeted for archaeometallurgical and archaeological surveys in 1981 as part of a program of lead isotope analyses. The broadly based survey focused on major regions of metal production and has been duly completed and integrated into a comprehensive data bank. The areas surveyed included Bolkardağ, Aladağ, and the Niğde Massif (Yener and Özbal 1987, Yener 1986, 1992, Yener *et al.* 1989a and b). The unique feature of this region as a mining district is its location near the strategic pass through the mountains, the Cilician Gates, and adjacent to the major artery through the mountains from central Anatolia to the Mediterranean Sea. Access southward from the mines and from the central Anatolian plateau to these immediate lowland areas are provided through these major passes. This region is integrally connected to the Levant in the south and is well known as a thoroughfare to the east (Alkım 1969), that is, to the Syrian and Mesopotamian heartlands. Most of our present information about the geographical distribution of mineral resources in Turkey stems from the Turkish Geological Research and Survey Institute (M.T.A.) and Etibank (the State Mining Institution), who extensively survey and operate the mineral reserves (M.T.A. 1964, 1970, 1972, 1984, English summaries in Ryan 1960 and de Jesus 1980; for earlier references, cautiously see Forbes 1963, 1964a and b).

The Bolkardağ Area

The central Taurus ores have often been described as polymetallic (Ayhan 1984) and the area has been identified as a highly complex geological zone (Akay and Uysal 1988). Iron is present at the 40% level as hematite or magnetite. Many of the ores are lead rich, in the range of 10 to 30% lead, and the lead is consistently accompanied by a high zinc content that runs at the 6 to 8% level, on average about one half of the lead concentration. The ratio of zinc to lead is about the same in slags as in the ores to which they relate. Copper is present up to 1.5% in some of the lead-rich ores and is found in a 6.7% concentration in mines within a few kilometers of lead-rich mines. Cobalt exists as high as 3.3% and tin as high as 0.3% in some outcropping veins. The mining region, therefore, could have been a source of copper as well as lead, tin, and silver.

The Bolkardağ valley is 15 kilometers long and lies about 50 kilometers north of the Mediterranean coast, northwest of the site of Tarsus. The major ore deposits at Bolkardağ are located on a 6-7 km horizontal extension and 550 m vertical width on the northern slopes (Ayhan 1984, Blumenthal 1956) (Fig. 6). Due to natural processes and mining activities, the Bolkardağ range is full of very irregular large caves, cavities, and tunnels and the form of mineralization in the region is quite unique. The primary sulfide ores are sphalerite, galena, and pyrite and there are massive secondary placer deposits of oxidized ores in the caves and cavities of the limestone mass. Some of these are layered like sedimentary deposits with gold content between 1-100 ppm and silver content at times higher than 6000 ppm, the majority falling between 100-1000 ppm. Since the deposits are quite soft and easily mined, it is suspected that the earliest mining activity in the region was simple panning with which metals such as gold could be easily recovered (Yener *et al.* 1989a and b). Porphyritic dikes are numerous and due to natural processes and mining activities the mountain range is full of galleries, some penetrating four kilometers into the mountain. Many of these show signs of having been worked in antiquity (Yener and Özbal 1987, 1989a). Similarly, by simple ore-dressing techniques, the other minerals present could have been concentrated prior to smelting. All the water and fuel necessary for such operations are available in the area in large quantities.

Ore samples containing high trace levels of a complex tin ore, stannite, were discovered on the steep slopes of Sulucadere at the crossing of two fault lines. The exposed vein of ore was in a pocket 110 cm by 20 cm along the fault line (Yener and Özbal 1987, Özbal and Ibar 1990) and mineralogical analysis identified it as stannite associated mainly with sphalerite, a zinc ore. The complex ore also contained galena, pyrite, arsenopyrite, pyrargyrite, and chalcopyrite. The elemental analysis yielded 33.1 ppm gold and 922 ppm silver. Covellite, chalcocite, limonite, malachite, azurite, anglesite, and cerussite are some of the secondary minerals which have been formed by the surface alteration of the primary ore minerals (Çağatay *et al.* 1989). The Sulucadere tin-bearing lead-zinc mineralization was formed in relation to the Horoz granodiorite, like the other known deposits of the Bolkardağ region. The hydrothermal solutions brought by these veins have followed the same route and formed the Sulucadere tin-bearing lead-zinc mineralization. Earlier references are also made to the occurrence of natural electrum (72.4% gold, 27.6% silver) and silver sulfides (argentite 87.1% silver) (Ladame 1938, Blumenthal 1956).

The Bolkardağ Area Site Survey

Very few archaeological sites in Turkey representing mining and smelting operations have been excavated and thus extractive and metallurgical methods in the metal-bearing zones have been difficult to reconstruct. Systematic, intensive, problem-oriented archaeological survey in these mountainous regions is a relatively newly applied technique because archaeologists seldom carry out large-scale surveys in areas with poor site definition and visibility. Survey in these areas involves hazardous mountain climbing and cold weather conditions in the summer, and often relies on local informants simply to make site discovery possible. Along with this, irregular terrain, small-sized activity areas, specialized production sites, and non-mound producing architecture all contribute to the sparse distribution of evidence. Nonetheless, village guides and local Geological Survey personnel have provided a wealth of basic archaeological information which has yet to be exhausted prior to the utilization of more advanced surveying techniques. The archaeological sites in the Bolkardağ mining district (Yener and Özbal 1987, Yener 1986) were undetected by previous travelers because their location on the slopes of the mountain range rendered them invisible due to the effects of erosion. The use of non-mound-producing wood for architecture, much like present-day mountain villages in the area, contributed to the virtual absence of the archaeology of mining sites until intensive surveys in the Taurus revealed their presence.

Fuel and charcoal production are the mainstays of mining and metallurgical industries. The burning of timber and its consequent use to smelt ore could have devastating effects on the forests of the region. Hamilton (1842), a nineteenth-century traveler to the Black Sea area, states that it took 260 tons of timber and 65 tons of charcoal in order to smelt 1.8 tons of argentiferous lead (galena 0.01-2% silver content). This resulted in 2.2 kg of gold and 15 kg of silver. Consequently, it takes vast quantities of wood to smelt the ore. The information gleaned through conversations with elderly miners at Bolkardağ revealed that the Ottoman smelters were located where forest resources were readily available and that the ore was carried to the smelters. As the resources of a particular slope were depleted, the smelter would then move to the opposite forested slope. This procedure was repeated twice in some miners' lifetimes; in other words, after about 40 years, the trees would be tall enough for the smelter to come back to his original slope. In this rather dynamic system, the timber resources were thus prevented from being totally depleted, and deforestation was relatively controlled—enough so that mining continued through the Ottoman period and provided most of the gold and silver for the palace in Istanbul.

The objective of the survey was to find all the archaeological sites within the catchment area of the Bolkardağ mines and to describe and sample them in sufficient detail to establish their size and archaeological phases. Thirty-three locations (B 1-B 33), 26 of which are settlements, were located within or very close to the mining district. In 1985, 8 more sites were found (B 34-B 37). In a minority of cases, site definition relied upon the recognition of archaeological mounding, cut features (ditches), or a scatter of pottery eroding down the slope. Anomalous finds of occupation in areas that, at the time, might not be expected to attract habitation on ecological grounds, provided a check on anticipated environmental adaptations in mining regions. A sequence of settlement and potential exploitation of the mine was derived from the earliest periods to its last known date of use, in the 1930s.

Seven sites (B 1, B 3, B 18, B 20, B 21, B 29, B 37), perhaps burials or mountain strongholds, were located on top of precipitous cliffs, which provide natural fortification with steep sides dropping 300 m to the river valley below. These cliff-top sites are roughly equidistant from each other (except for B 37), about 500 meters apart, and lead into the valley proper where gentler slopes provide wider space for mounded settlement formations. It is in these flatter surfaces or intermontane valleys that mounds such as Iron Age Porsuk, the Medieval site Gümüşköy (B 16), and Chalcolithic/ Early Bronze Age Garyanın Taşı (B 26) are situated.

There are two major slag deposits in the Bolkardağ region: the Madenköy slag mound (B 7) is estimated to be approximately 96,000 tons and the Gümüş slag deposit (B 16) is estimated to be about 620,000 tons (M.T.A. 1972). Besides these major sources located by the Turkish Geological Survey, scattered slag samples were identified in archaeological survey on the northern slopes of the Maden valley, especially around Yediharmantepe (B 5), Katırgediği (B 19), Pancarcı Kale (B 20), Tavşanın Yeri (B 25), and Geyik Pınarı (B 21). Evidence of furnace structures and refractory materials were also observed in the same locations. One such location (B 5) had the remains of several furnaces lining both banks of a dried-out stream, facing the prevailing winds. These round structures, measuring roughly 1.5 m in diameter, were associated with a scatter of slag and pottery. There was even a pot-bellows nozzle in situ in the wall of one furnace (Yener and Özbal 1987). The major slag deposits at Madenköy and Gümüş were believed to date to the Classical Greek, Roman, Byzantine, and Ottoman periods (Yener and Özbal 1987, Yener and Toydemir 1993, Yener *et al.* 1989a, Yener 1986). However, archaeological surface surveys have indicated the presence of pre-Classical sites in the proximity as well. The average elemental analyses of these four major groups of slag (Madenköy B 7; Gümüş B 16; Yediharmantepe and surrounding areas B 5) are listed in Table 1. It is surprising to find considerable concentrations of gold and silver in some of the slag. The average tin concentration in the slag is even higher than that seen in the ores. In fact, 24 of the 29 samples contained tin at about 1540 ppm, suggesting that in the later periods tin was not the targeted metal. These results suggest that tin ores were located not far from these smelting sites.

The Çamardı Area

The tin mining complexes (including Kestel) and their associated specialized activity areas are situated upslope from several rivers coursing through the Niğde Massif, a large volcanic dome formation 40 km to the north of Bolkardağ. Located in the central Taurus mountain range, the mines are 4 km west of Çamardı and the village of Celaller, Niğde province, and 80 km north of Tarsus. They are strategically situated along the north-south Ecemiş fault zone, providing access both to central Anatolia to the north and to the Cilician plains and the Mediterranean coast to the south, passing by the Bolkardağ valley. Streams have cut deep valleys at the northern side of the fault and have yielded placer-rich alluvium. Two streams, Kuruçay in the west and Burçdere in the east, drain the tin-gold anomaly zone.

Plate tectonic activity is quite intensive in the Celaller area as revealed by a succession of subduction zones with extensive mineralization. The Niğde Massif has a gently rolling terrain with outcroppings of diabase, granitic material, and marble as part of dolomitic limestone. The tin mineralization, cassiterite, occurs within the granite and also along the granite borders. Hematite-bearing quartz veins, pegmatites, and tourmaline-bearing quartz veins are abundant along the tin mineralization. Many veins of different elemental and mineral composition occur at Kestel and these contribute to its alluvial deposits, including scheelite, cinnabar, apatite, pyrite, pyrotine, rutile, titanite, monozite, and gold (Çağatay and Pehlivan 1988, Çevikbaş and Öztunalı 1991). Cassiterite of several different colors—burgundy, red, orange, yellow, the more common gray/black—reflects a variety of different trace impurities that occurred in the Kuruçay stream (Plate 2). The Kestel mining complex, which includes a number of galleries such as Kestel, Sarıtuzla, Mine Damı, and Sulu Mağra, was cut into a slope composed of granite, marble, gneiss, and quartzite 200 m above Kuruçay stream (Yener *et al.* 1989b, Kaptan 1995a and b, 1989). These and other mines with collapsed entrances surveyed along the streams yielded significant amounts of cassiterite.

At Kestel mine, cassiterite ore was also extracted from thin cracks within the marble, limestone, and quartz schist matrix (Willies 1993). Tin also occurs in conjunction with hematite and manganese oxide which would have replaced the marble in spaces. Large, empty domes and semicircular domes in the mine are evidence of total extraction of the ores in these spaces. Quartz and tourmaline veins and poor value hematite veins were left intact (Willies 1990, 1991, 1992). High-grade iron ore was pecked off and discarded, as evidenced by the large, unused quantities along the slope talus debris. Small pegmatite veins were found in a nearby fault, which may have acted as a "mineralizing fluid feeder," although pegmatites were worked in the wider area a few kilometers around the mine (Willies 1995). The possibility that gold was being mined also exists since the mine was found during geochemical sampling of placers by the M.T.A. This was considered at the very outset of the analysis of the materials from Kestel (Earl and Özbal 1996). Tin is usually found in the same general area as gold and this is certainly true, for example, in Cornwall. Cornish gold is as well-known as Cornish tin. The Turkish Geological Institute view is that the mine is indeed a tin prospect although the region is recognized as an important gold source.

The Çamardı Area Site Survey

Most research into the technology of metal production has concentrated on assemblages excavated from the lowland urban agricultural sites in Cilicia, central Anatolia, Syria, and Mesopotamia, all major consumers of metal products. On the other hand, information from specialized function sites in the resource zones has been comparatively scarce, leading to a perspective on metallurgical techniques skewed toward the consumers. Recognizable tin sources in the eastern Mediterranean have been few up to the present and the economic and technological significance of a tin mine in this region has never been assessed archaeologically. The increased demands for raw materials on local Anatolian industries could hypothetically demonstrate a heightened usage of bronze. Such phenomena should have archaeologically observable correlates. In the metal source zones such as Çamardı, the relationship of overland trade to settlement history should reflect the increased external demand for metals in the Early Bronze Age. Establishing the number of sites in the principal north-south passes through the Taurus, typically at locations of considerable strategic importance, was therefore critical.

The 1988 archaeological survey was centered at the Kestel tin mine and aimed at the recovery of the patterns of past human activity and settlement within a mining district of 40 km^2. The survey explored environmental and technological factors that might have conditioned metallurgical and habitation site locations; these were proximity to ore sources and fuel, accessibility of passes through the mountains, and the location of rare agriculturally fertile intermontane valleys. The survey methods used were transect and circular sampling, input from area specialists of the Turkish Geological Survey, and local informants. These strategies aided in the location of workshops, mines, and settlements in difficult topographical terrain (Fig. 7). Thirty-three sites were mapped in this fault zone during the 1987 and 1988 surveys (Fig. 8 and Table 2). The sites were labeled Ç1-Ç33 and located on maps obtained from the Turkish Geological Survey (1:25,000 no. M33 ba Kozan). Mounded sites were found as well as metallurgical installations, specialized function sites, and workshops not previously detected.[4]

The surveyed sites are in three basic locations: along routes at strategic points, in alluvial areas suitable for agriculture, and on hilltops. The majority of the 33 sites line the passes through the Taurus mountains. They are strategically located at the crossroads of two major routes: 1) the well-known silk route from Cilicia and the Mediterranean basin which winds northward through the Taurus passes along the Ecemiş fault directly to Kayseri in central Anatolia (site numbers Ç3-8), and 2) the turn-off of the silk route to Niğde and the northwest (site numbers Ç13, 17, 23, 30, 31). Some of the sites (site numbers Ç6-8, 10-12, 24-28) are also located within the alluvial plains of major rivers flowing through the passes. These primarily mounded settlements could have been agricultural subsistence bases for the specialized operations at Göltepe and Kestel mine, situated in areas less suited for farming. Several sites (site numbers Ç4, 27-30) are located on the summits of high hills with a commanding panoramic view of the valley below. It is important to note that, with perhaps a few exceptions, every hilltop and mountainous peak in this region had a site on it. Extrapolating from this to the rest of Turkey and considering the fact that most of the country is mountainous, the potential magnitude of populations unaccounted for in the archaeological record is enormous.

In addition to these wider survey aims, two intensive surface investigations targeted more specific site-oriented goals (Yener 1989, 1990). The results of these and the archaeological probes at two of these sites (Ç2 and Ç13, Kestel and Göltepe, respectively) are reported below. The specific aim of the intensive investigation was to illuminate the morphological structures of both sites in preparation for future excavation, that is, to map out the seemingly promising points based on surface

---

[4] More comprehensive reports will be published upon fine tuning of the cultural and chronological indicators in the pottery, especially the Early Bronze Age sequences of Göltepe.

indications. In addition the investigation explored the possible functional relationship between Göltepe and Kestel mine. The contemporaneity of both sites had to be established in order to eventually investigate the organization of an extraction, production, and habitation system. The inferences drawn from the survey led to the decision to excavate the sites of Göltepe and Kestel. The most important aspects of these investigations have been the light shed on craft specialization in a metal-rich zone, the context of metallurgical innovations, and the possibility that tin metal was being provisioned to the urban polities of the Early Bronze Age from this area.

Kestel mine (Ç2) is located above the Kuruçay stream two kilometers west of Celaller village (Plate 3). Although the Niğde Massif has by no means been exhaustively surveyed and settlement data collected around the polymetallic source area of Çamardı and Bolkardağ has just begun to be processed (Yener *et al.* 1989a and b, Aksoy 1998, Aksoy and Duprés *in prep*), several trends are beginning to appear. Göltepe is by far the largest Early Bronze Age site and is located closest to the Kestel mine complex on marginal agricultural land. Third and second millennia B.C. sites within ten kilometers of the mine (site numbers Ç10, 25, 28) are mostly an average of 1-5 hectares in size and are located in the more fertile agricultural river valleys. A number of mounded sites exist along the critical Ecemiş fault zone coeval with the specialized site of Göltepe. Although situated on more arable land, the sites in the passes also processed metal, judging from the metallurgical debris found on their surface.

Göltepe is located in an intermontane, relatively fertile pocket of land 4 km from the major passes through the Taurus mountains (Yener 1992, 1993, 1994a and b, 1995a-c, 1996a, 1995d). It is assumed that Göltepe was a specialized metal processing site, and was agriculturally self-sufficient. Modern land use in this region today indicates that legumes, fruits, and wheat can be intensively grown in agriculturally fertile subzones, while the upland slopes provide transhumant populations with pasturage. This concentrated pastoral productivity is the mainstay of the region today. If Göltepe-Kestel was relatively self-sufficient in terms of agricultural production, an alternative to, or a stabilizing factor for, this self-sufficiency may have been a provisioning system where foodstuffs from local agricultural sites were exchanged for metals from special production sites. However, it is also possible that the mining and processing sites were entirely seasonal. Seasonal mining strategies such as these were documented by 19th-century travelers to the mines in the Black Sea area of northern Turkey (Hamilton 1842). In these instances, the miners were transhumant pastoralists, who mined part-time during the summers while in the highland pastures and then returned to their lowland farms at the end of the season. The semi-nomadic links with metallurgy have been pointed out in other areas of Anatolia as well as in Russia (Cribb 1991, Chernykh 1992). The supply of subsistence goods may have linked into a prevailing system of transhumance (Bates and Lees 1977).

The influence of Alpine ecology on the early development of metals could be profoundly associated with seasonal migrations and thus would have provided the mechnism for transporting semi-processed materials to the lowland areas of Cilicia and the Amuq. Earlier hypotheses have pointed out the inherent mobile nature of metalworkers (Childe 1944) and the possible nomadic porters of metals (Crawford 1974). The village of Celaller provides a rich ethnographic example of on-going transhumance practices. Originally Yörük nomads migrating from the lowland Cilicia and Syrian coastal littoral, the local population was settled into the present village when the border between Turkey and Syria was established prior to WW II. Within the central Taurus, the Niğde Massif area (1600-2000 m altitude) was originally the summer pasturage of these nomads and when given a choice of land, they chose an area more conducive to their livelihood of camels and herds of sheep and goat. Their economy today still relies on pastoralism, carpet weaving, and limited agriculture. The village owns vast hectares of pasture lands in the Niğde Massif and continues to migrate further upland every year thus continuing the transhumance legacy. It is important to note that this is a local pattern that was adapted regionally and is carried out by a splinter segment of the society, the women. For six months out of the year, the women of the village take a few children and go upland to the higher elevations (2000-2500 m) with their herds. The men generally stay in the village and work on the meager agriculture. The highland dairy industry run by the women consists of making yogurt, cheese, and dairy products and shearing the sheep for eventual use in the carpet industry which occupies them during the winter months.

It is not surprising to see that Göltepe was integrated into a network of settlements between the Taurus at points of strategic importance to routes leading to lowland sites during peak periods of metal demand. Indeed, given the economic and technological significance of tin, an increase in aggregate settlement size and quantity in the metal-producing zones is apparent, as is the placement of sites at strategic crossroads, especially in the third millennium B.C. That is, according to transit models, the relationship of overland trade to settlement history (Steponaitas 1981) at metal source zones reflects an increased external demand for metal in the Early Bronze Age. A number of urban sites are located across the principal north-south passes through the Taurus, typically at locations of considerable strategic importance. The impact of the increased demand on the highland producers is detectable in the establishment of large-scale production sites and population growth in the highland settlements. These sites not only controlled the flow of intermontane traffic, but were themselves specialized metal manufacturing sites. This inference is borne out by the presence of large mounded sites coeval with the fortified first-tier production site of Göltepe along the critical Ecemiş fault zone. The Cilician Gates, an important pass through the central Taurus range, provides access from the Mediterranean Sea coast to important urban settlements in the central Anatolian plateau. Both the quantities of material moved and the multitiered aspect of this industry should be factored into any exchange reconstruction.

The Kestel Intensive Surface Survey

A separate survey operation targeted the distributions and densities of artifacts on the mining slope, the chronological range of surface materials, and the nature of activities carried out at the entrance location (Fig. 9). A preliminary inspection of the Sarıtuzla slope debris, the tailings from the Kestel mine, and the open pit and collapsed mine entrances indicated certain dense concentrations of Bronze Age finds with definable spatial limits. It was believed that these distributions would help establish overall man-mine relationships in the third millennium B.C. Accurate estimates of tailing size are important for two reasons. First, relatively precise estimates of the shape of activity loci pertaining to mining and workshop activities at the mine needed to be made. Second, precise measures of the artifact densities might prove valuable in locating collapsed, nonvisible mine entrances, workshop sites, or domestic quarters of the miners.

The standard procedures involved laying out a grid with a site datum near the geographical center of the site, Kestel mine (Fig. 10). The datum was located 15 m north of the main Kestel mine entrance at an altitude of 1878 m. The Sarıtuzla slope was mapped with 30 grid-squares, 50 meters square in size in a one-kilometer area. Each grid-square was divided into four equal triangles measuring 625 m^2 to provide better control over collection and recording. All intact groundstone tools and sherds were collected by walking in a N-S or E-W direction, each participant separated from the next by 2 meters. The diabase and gabbro stones, which are not local to the Kestel-Sarıtuzla slope, were counted and left on site. The densities of artifacts and structural features were mapped for each grid-square.

Several observations resulted from this surface survey. It was apparent that at the datum and one other area to the east, the distribution of surface artifacts were coincident with the location of stationary ore-crushing installations on marble bedrock and mine entrances. Within the tract itself, it appeared that stone hammers and other groundstone tool densities were largely restricted to the outcropping granite and quartzite zones. Ore-dressing activities were primarily located near mine entrances. Evidence of pecking, crushing, and more rarely grinding was characterized by clusters of small, circular, mortar-like depressions in the bedrock, which were identified as stationary ore-crushing areas. Distinguished by high artifact concentrations, activity loci were prominently situated on the Kestel slope in three main areas. These were designated Activity Loci A-C based on the use of the marble platform rock as mortars. Located mostly at the same elevation as Kestel mine, other platform rocks farther down the slope also showed traces of shallow hollows, although erosion made them more difficult to discern, and may indicate older workings nearer the stream. One major installation located on the roof of the Kestel entrance was mapped in detail (Fig. 11). Utilized extensively for ore crushing, the marble surface had 216 hollows, ranging in size from 5-9 cm in diameter and 1-4 cm deep (Kaptan 1989: Fig. 2). The discrepancies in size seen in the diameters and the depths of these hollows may be functionally related. That is, when the hollow became too deep to function as a mortar, it was abandoned for another flat surface.

The most abundant surface finds, other than the pottery, were groundstone tools. Certain typologically distinct ore-processing equipment emerged from the surface surveys. The principal materials used for the groundstone tools were minerals known for their hardness, such as gabbro, andesite, and diabase. These tools appear to be functionally related to ore processing. Other tools were fabricated from marble and quartzite, while sandstone was utilized for molds (Kaptan 1990a, Hard and Yener 1991). The stone tools are primarily small handstones (Kaptan 1990: no. 7) and larger stationary tools. Some grinding is indicated by flat surfaces. The tools have small to medium circular hollows, large concave ground surfaces on both portable and larger non-portable stones, and battered surfaces. A large multifaceted diabase ore-processing tool, 30 x 90 x 30 cm (Kaptan 1990: nos. 1, 2), was found 50 meters east of the Kestel mine entrance. The obverse had ten hollows and was used as an anvil. The reverse was concave and was probably used as a quern for grinding purposes.

Another category of groundstone tool was the vesicular basalt saddle quern (Fig. 12: K), a type which is more often attributed to domestic use. These were found adjacent to prominent, stationary, ore-crushing installations. Excavated examples have been found in a number of lowland sites in southwestern Asia and have a range of dates from the Neolithic period through the Late Bronze Age (Goldman 1956: Fig. 419: no. 113). The higher densities of groundstone tools, especially ones with battered surfaces, were centered around the outcropping ore veins where several collapsed mine entrances could be discerned. The tools were assumed to have been used to process either ore or food. With the knowledge gained from comparable work at Göltepe, the data collected on the surface of the site could be used to delineate specialized activity areas relating to mining and processing as opposed to domestic activities.

Most of the pottery was dateable to the Early Bronze Age—and rarely to the Chalcolithic and Byzantine periods—and was distributed in a pattern similar to that of the stone tools. The phases of the Early Bronze Age pottery found at Kestel were not differentiated until Göltepe was excavated in 1990. The ceramic typology at Kestel falls primarily into two periods, the whole extent of the Early Bronze Age and the Medieval period. The first is characterized by a dark red or black burnished tradition, some pieces with micaceous temper, a hard-fired clinky metallic ware, an orange gritty ware, coarse chaff wares, and crucible fragments (Plate 4). Crucible fragments found on Kestel slope were analyzed by SEM and contained high tin assays. Painted (Fig. 14: Q) and buff Chalcolithic sherds at Kestel seem to indicate an earlier presence on the Kestel slope. Aside from the ceramic and groundstone finds, several grid squares had what appeared to be talus debris, obsidian tools (Fig. 12: E), metal objects (Fig. 12: F), and an animal figurine (Fig. 12: A); these correlated well with the densities of other artifacts.

Sounding S.B.

In order to understand the chronological relationship of the surface remains to mining and the relationship of soundings to slope material, a probe (S.B.) was placed in the area of densest cultural debris on the slope (Fig. 9). Sounding S.B., measuring 1 x 2 m was placed outside the entrance of Kestel mine close to the marble platform rock used for ore dressing. The location was selected not only to date the earlier phases of mine tailings and ore crushing, but also to identify the ore being processed. Samples of soil were taken every 10 cm and after about 40 cm of slope talus, stratified cultural deposits emerged. Pottery, vitrified structural mud lumps with branch impressions, bucking tools (Fig. 13: F), ore nodules, and charcoal, as well as bones, emerged in sequence. After -70 cm an ashy, charcoal-laden horizon emerged underlying the vitrified mud fill above. The earliest locus, representing a collapse level at -93 cm, yielded a slab of burnt structural mud, plastered on one surface, suggesting the presence of architectural units outside the mine.

The ceramics were primarily dark burnished (Fig. 14: B, C), clinky metallic, and orange gritty wares. Large storage vessel rim fragments, micaceous unfinished ware, plain simple wares, coarse wares, chaff-faced ware, crucible fragments, and what appear to be mid-third millennium B.C. Syrian metallic ware (Kühne 1976) characterized the basal units. The probe ended at a depth of -1.40 m upon reaching bedrock. The results suggest that architectural units were located in proximity to mine entrances. These structures may have been workshop quarters for the miners and need to be fully exposed by excavation. The contemporaneity of the slope sounding to the mine soundings suggest that certain ore-dressing as well as habitation functions were localized. The small numbers of painted Chalcolithic and straw-tempered hole mouth jars of a Late Chalcolithic type may indicate an earlier presence in the vicinity of the mine itself.

Excavations at Kestel Tin Mine

A separate phase of the operation independent of the initial regional field walking in the Çamardı area involved returning to the Kestel slope location (Sarıtuzla) for more detailed study. The focus of the research strategy employed at this stage was determined by the problems to be investigated. More information was needed on the extent of the tin mineralization, the nature of the activities carried out in the mine, the size of the artifact concentrations inside the mine, and the chronological range of materials.

Cassiterite (tin oxide) was found at Kestel mine by the M.T.A. (Pehlivan and Alpan 1986, Çağatay and Pehlivan 1988). Recent analyses by atomic absorption spectroscopy of ore from veins remaining unmined inside Kestel mine indicated that after extensive mining in antiquity the veins still contained up to 1.5% tin. The results of the ore analyses have raised questions about the accumulation of artifacts and ore-crushing features near the entrances of this mine. Kestel mine was initially examined between 1987 and 1989 when four 1 x 2 meter soundings (S.1-S.4) were placed inside the galleries and one was placed at a workshop adjacent to the entrance. The dating of the operations at Kestel mine relied heavily on radiocarbon dates and stylistic studies of ceramics. Early Bronze Age (ca. 3200-2000 B.C.) sherds as well as charcoal, bones, and groundstone tools were recorded inside and outside the mine, and architectural daub fragments emerged from the workshop sounding. A preliminary sketch of Kestel mine by the Turkish Geological Survey was published (Yener *et al.* 1989a) and a more detailed, revised map has now been drawn (Fig. 15). The mine was initially divided into eight chambers and numbered with Roman numerals I-VIII. These loci represent the extent of the mine accessible prior to a clearing operation in 1991 which unblocked shaft debris leading to a vast downslope gallery complex. The eight divisions are not only a recording device, but also roughly divide the mine on the basis of the morphological differences in the apparent techniques of mining. From 1990-1996, new soundings by a collaborating U.K. mining specialist team expanded knowledge of the extent of the mine (Willies 1991-1995, Andrews 1994, Craddock 1995, Yener 1996, 1997a and b).

Certain visual clues suggest that the methods of extraction were different from period to period. Fire setting and hammering with large groundstone battering rams were the main methods of extraction. The earlier workings, found primarily in the northwestern sector of the site, are predominantly fire set and very small-scale operations compared to the subsequent mining events. Later workings, which cut through the earlier ones, are larger in scope and display evidence of both fire setting and heavy hammering, perhaps indicating improved mining techniques. The smaller and seemingly earlier tunnel-shaped workings (Chambers III-VII) could be contrasted with the much enlarged entrance area (Chamber I) incorporating a large chamber and central pillar (Chamber II). The earlier adits, measuring 60 cm in diameter, were cut into the limestone and generally led upslope at an angle of approximately 30′ (see Chamber VI). The limestone walls are smooth faced and curvilinear with an appearance resembling erosion by water. No signs of battering were seen on the interior face of the adits, but dome-like fire setting features (Willies 1990, 1991, 1994) were apparent on the roof, and along the floor levels of this part of the mine. The fire setting method of extracting ore entails lighting a fire under a vein, and then quenching the super-heated walls with water, causing them to crack (Craddock 1985, 1995). The ore is next cobbled with a hard-stone tool made from diabase and removed. Since the limestone is interlaced with mineral-filled microfractures, which naturally cleave along curvilinear lines (Bryan Earl, personal communication), dome-shaped alcoves result from this fire setting process and thus are an indicator of possible Bronze Age mining.

Radiocarbon dates obtained for the Kestel mine workings establish it as the oldest tin mine found to date (Table 3; radiocarbon 2 sigma calibrated dates range from 3700-2133 B.C.). A number of early shaft and gallery complexes worked with stone hammers in neighboring countries show similarities. Rudna Glava in Yugoslavia and open-pit mining of complex copper ores at Ai Bunar in Bulgaria (Jovanovic 1978: 9-10, 1980, Jovanovic and Ottoway 1976, Chernykh 1992), Timna, Israel (Rothenberg 1990), and Fenan, Jordan (A. Hauptmann 1995) have yielded important information about comparable mining technologies for copper in the Chalcolithic period. In Turkey, similar methods are found at the Black Sea site of Murgul in the fourth millennium B.C. (A. Hauptmann 1989, Hauptmann *et al.* 1992, Lutz *et al.* 1994); radiocarbon corrected dates

3340-3040 B.C. and 3635-3495 B.C.) and at a mine near Tokat-Erbaa dating to the fourth-third millennia (Kaptan 1986, 1990); similar stone mining tools were found at a silver mine in Kütahya (Kaptan 1984). The Murgul information is especially interesting because of the analysis of its cake-shaped slag.

Comparative sequences for mining techniques based on a typology of structural differences in the galleries (Craddock 1985) provided a working hypothesis that some extraction cavities represented different periods of mining. That is, the smaller, narrower fire-set galleries represented an earlier phase, while the larger chambers were thought to be later re-excavations. This conjecture was supported by the evident gutting and cross-cutting of smaller galleries by the larger Chamber III, leaving only small segments of earlier workings; this is visible in the cross cutting between the pillar and wall at survey station 2. A span of use for the mine complex beyond the Bronze Age is supported by ceramics found on the surface and in soundings inside several chambers.

A sounding in Chamber I, the closest to the entrance, was investigated in 1987 and the surface scatter was mostly Medieval and Early Bronze dark burnished and plain simple wares (see below for descriptions and parallels). Chamber II yielded early coarse and dark burnished wares. Medieval wares, as well as a Byzantine coin, were also found in Chamber V and Chamber VIII. The latter also contained a large, diabase, large-rilled mining pick, weighing 5.3 kg (Kaptan 1990: Fig. 24: 1, 2) (Fig. 13: I), which was probably hafted using bent branches around the groove in the center. Other finds included a bucking stone (Fig. 13: H), and some Early Bronze micaceous unfinished (Fig. 14: U) and orange gritty ware types. Several diabase hammerstones, some with grinding surfaces and hollows, suggest that ore dressing was also taking place inside the mine in the Bronze Age.[5]

Kestel Mine Soundings S.1-S.4

Finding ceramics of several periods on the surface of the galleries and the divergent extraction techniques strengthened the notion that ore was being removed over several periods and that a chronological sequence of mining might be reconstructable. In order to test the assumption that the mine was exploited over a long span of time and that cassiterite was the targeted mineral, four small-scale 1 x 2 m soundings (S.1-S.4) were initiated inside galleries II, III, VI, and VII. These soundings were dug in arbitrary 10 cm levels to obtain soil and charcoal samples for mineralogical and radiocarbon

---

[5] Comparable examples dating to the Chalcolithic and Early Bronze Age were found in Fenan, Jordan (Hauptmann, Weisgerber, and Bachmann 1989), Timna, Israel (Rothenberg 1990), Rudna Glava, Yugoslavia (Jovanovic 1978), and Kythnos in the Aegean (Gale *et al*. 1985).

analysis. The assumption was that the detritus of mining activity would yield important information about the original ore-body composition and would help date the mine. Geochemical and mineralogical analyses were performed on samples taken from these strata to determine whether mining in fact took place, whether the gallery was used for shelter, and which mineral was being extracted. SEM and X-ray diffraction analyses in 1987 of soil accumulated in the mine and ore veins indicated that the ore being worked in the mine entrance was tin oxide or cassiterite. The radiocarbon results demonstrated that the mine was utilized during the third millennium B.C.

Sounding S.1 (1.5 x 1 m) was placed in Chamber II, adjacent to a stone pillar 3.7 m high left by ancient miners to hold up the roof of the mine (Fig. 15). This location was chosen because of its proximity to the main entrance. Rich veins of hematite measuring 58 cm thick and 2.7 m long were left unmined and can be seen on the surface of the ceiling, suggesting that the miners were not after the iron ores. Surface finds included a lamp fragment with an oily black soot coating its interior (Fig. 12: B), a discoid diabase tool, and two hammerstones. This sounding yielded a consistently mixed deposit of Early Bronze Age and Byzantine sherds and fragments of iron artifacts. Radiocarbon dates from charcoal samples from a depth of 60 cm in this sounding, the floor of the chamber, yielded a date of A.D. 380 ± 60 (calibrated A.D. 347-609, 2 sigma). The bones of various animal species, domestic goat (*Capra hircus*), a canid, probably dog (*Canis sp.*), and camel (*Camelus sp.*), were also identified in this sounding. The mixed nature of the debris containing ceramics from the Bronze Age to the Medieval suggested that activity at this spot near the entrance represented a complex series of extraction episodes and/or later use as shelter. The presence of cassiterite in the soil strengthened the assumption of tin mining.

In 1987, soundings S.2a and S.2b initially tested the gallery floor deposit in Chamber VI (Plate 5a) for mineral samples (Plate 5b). This was subsequently expanded in 1988 for datable material with a trench measuring 1.5 x 1 m placed at the confluence of five upsloping galleries, some of which measured a scant 60 cm in diameter. These galleries, with circular cross-sections, differed morphologically from entrance Chambers I and II. The rationale for putting soundings S.2a and S.2b in this part of the mine was twofold. First, the gallery features were assumed to represent an older mining activity complex than the Chamber II workings. Second, the more primitive extraction technique of this chamber suggested that it might yield in situ evidence for earlier phases of exploitation.

The sounding yielded a mixed deposit of Medieval and Bronze Age sherds, a glass bracelet fragment, and diabase tools in the first 40 cm (Loci

1-4). The Early Bronze Age sherds were mostly dark burnished and unburnished varieties (Fig. 14: L, M, P), red burnished (Fig. 14: K), and micaceous finished (Fig. 14: O). A pink ware was dated to Karum IV levels (Karum IV, III c. 2000 B.C. by Middle Chronology) at Kültepe,[6] suggesting tantalizing connections to early second millennium central Anatolian sites. Below a depth of 40 cm, the pottery became more homogenous with a dark, highly polished Early Bronze Age ware predominating. Some cruder examples such as a hole-mouth jar and several straw-tempered types also emerged in the lowest strata. A circular hearth framed by a clay perimeter emerged at 40 cm, locus 4, full of charcoal and ash in a dark flaky soil strata. Ceramics of both the highly burnished black and red ware (Fig. 14: G, I) and coarse ware varieties (Fig. 14: S) were abundant at this depth. Some sherds had burnt exterior surfaces suggesting food preparation inside the mine galleries.

Below the -45 cm depth, the pottery became more homogenous with a dark, highly polished ware predominating (Plate 6b). Some cruder examples such as a hole-mouth jar and several straw tempered types (Plate 6a) also emerged, suggesting a Late Chalcolithic phase somewhere in this mine as well. It is also possible that the Chalcolithic pottery slid into this gallery from earlier open-pit mining operations situated 50 meters upslope. Open-cast mining is generally thought to precede the shaft and gallery systems and thus the Chalcolithic pottery may have slipped into this spot through the vertical shaft that was emplaced after the extraction pit operations. However, a precise Late Chalcolithic dating for this pottery must await the study of comparable sequences (Summers 1991) elsewhere in the Çamardı and Niğde area when they are excavated. A layer of collapse was reached at -60 cm bringing up the possibility that a blocked vertical gallery existed at this spot. The sounding was therefore stopped at -93 cm for safety reasons. This basal unit layer of collapse from -60 to -93 cm has a massive character and suggests spoil from mining activity.

The heterogeneous nature of the pottery in the loci of sounding S.2 emphasized the necessity of having prior knowledge of mining techniques before selecting soundings inside a mine. In the case of Chamber VI, the upper 30 cm of sounding S.2 yielded a layer of mixed Bronze Age and Byzantine sherds that clearly resulted from downslope sliding of mining debris from upper story operations. Material dating to early mining events derived from higher elevation galleries which had spilled debris during modern erosional episodes and created a reverse stratigraphy in the gallery below where sounding S.2 was placed. At a greater depth in the sounding, less mixing was apparent, but small amounts of later wares continued until

---

[6] I thank Aliye Özten for her identification of this ware and its similarities to Karum IV materials.

at least the -40 cm depth. Geochemical evidence showed that "the laminations in this depth suggest a water-deposited, non-cultural phase during which mining activity was more remote from sounding S.2."[7] For this reason, only samples below 40 cm were utilized for radiocarbon dating and mineral identification. The radiocarbon results on samples of charcoal yielded dates of 2070 ± 80 B.C. (calibrated 2 sigma, Struiver and Pearson curves, 2874-2350 B.C.), 1945 ± 70 B.C. (calibrated 2576-2147 B.C.), and 1880 ± 65 B.C. (calibrated 2469-2133 B.C.) and suggested that locus 5 at -68 cm could be dated to the third millennium B.C.

The discovery of a large diabase mortar or anvil (Fig. 13: D) with 2 circular hollows on one surface (Kaptan 1989: Fig. 3) provided information about the specific tools of extraction and beneficiation in this chamber. Other surface finds in Chamber VI included a lamp, a diabase pestle (Fig. 13: E), and Early Bronze Age dark burnished pottery (Fig. 14: F, J, N). Large amounts of faunal material,[8] some of large-hoofed animals—red deer (*Cervus elaphus*), ass, or horse (*Equus sp.*)—as well as small birds, rodents, hyena (*Hyaena sp.*), and tortoise shells, were also found on the surface (Yener *et al.* 1989b). The bones of various animal species were recovered from the -68 cm level, and are representative of all the loci in this sounding. They include domestic goat or sheep (*Capra hircus, Ovis aries*), an ungulate (probably *Bos*), dog or other canid (*Canis sp.*), pig (*sus scrofa*), bird, and a rodent. Macrobotanical samples yielded the remnants of mallow (*Malva sp.*), oak (*Quercus sp.*), coniferous wood, probably either fir (*Abies sp.*) or juniper (*Juniperus sp.*), and possibly almond (*Prunus sp.*) (Yener *et al.* 1989b).

Sounding S.2 has revealed several discernible features about this chamber. First of all, the mine dates to the third millennium B.C. and contains cassiterite deposits. The technique entailed fire setting, then battering the ore with heavy hammerstones. Mortars, pestles, and bucking stones indicate that some ore beneficiation was also taking place inside the mine. Secondly, the presence of pottery with open forms, the domestic fauna, and a hearth suggest a certain amount of eating was done inside the mine. Larger ceramic forms were perhaps for storage, presumably to contain water or foodstuffs.

Sounding S.3 (2.0 x 0.75 m) was placed in Chamber III. This gallery could be entered from Chamber II by crawling through a narrow entrance 85 x 112 cm or from a narrow adit off the main entrance. Even though the larger dimensions of Chamber III and the technique of mining seemed morphologically different from the earlier Chamber IV, a familiar early

---

[7] Robin Burgess report July 24, 1991.

[8] I thank Ibrahim Tekkaya from the M.T.A., Ankara for the faunal report.

feature of a dome-like cut in the limestone, attributable to fire setting, was visible on one of the walls. The purpose of this sounding then was to date this chamber, which was different in its size and features from the previous chambers that were investigated. After clearing the black organic surface, locus 1 (0-10 cm), which consisted of bones, beetle parts, and debris, the trench was dug at 10 cm arbitrary levels labeled Loci 1-5 from the surface. The cross-section revealed six distinguishable stratigraphic levels. Although all levels revealed some amount of mixing, charcoal, ceramics, and bone started emerging in more coherent stratigraphic layers after 10 cm. A mixed layer of mining and cultural debris comprises locus 2 (-10-20 cm). Underlying this at -20 cm is a thin, dark gray horizon with charcoal flecks, bone, and ceramics. Locus 3 (-20-30 cm) also contains a light brown clay layer with charcoal and locus 4 (-30-40 cm) is a burnt layer with charcoal, large bones, and ceramics. The lowest, locus 5 (-40-65 cm), directly on the gallery floor, is a clay deposit with charcoal flecks.

Examples of ceramic types from the surface, locus 1, were mostly a mixture of Medieval (Fig. 14: A), Iron Age (Fig. 14: R), and Early Bronze Age wares. The latter were orange gritty, micaceous unfinished (Fig. 14: T), dark burnished, and fine slipped wares which persisted into locus 3. As with sounding S.2, an early second millennium pink ware paralleled in Karum IV levels at Kültepe and a micaceous slipped variety (Fig 14: H) were also present into locus 4. Burnished wares from locus 5 resembled the Early Bronze Age red-black wares of sounding S.2.

The identification of the faunal specimens from this sounding reiterated the presence of species prevalent in the other soundings: domestic goat, an ungulate, dog, and bird were noted from all levels. The teeth and vertebrae of a horse or donkey (*Equus sp.*) was the only difference and came from locus 2. Large antlers found in the mine may have been used as picks (Plate 7). The stratigraphic and geochemical evidence from this sounding suggests that there are at least two distinct episodes of relatively intense cultural activity. Loci 1, 2, 4, and 5, especially, appear largely undisturbed. The finds suggest a long period of activity with modifications in ore extraction methods coming in after the Early Bronze Age.

Sounding S.4 (2 x 1 m) was placed in small extraction gallery VII. The smooth-faced walls and other signs of fire setting suggested an early date for this chamber, hence providing a cross-check with the data of soundings S.1-3. Unfortunately the deposit was only 20 cm deep and after samples were taken from the floor of the mine, digging was abandoned when the gallery floor was reached. The finds from this gallery include predominantly Medieval wares (Fig. 14: D, E) and some micaceous finished and unfinished wares dated to the Early Bronze Age.

Soundings in Kestel mine entrance area were continued by the Historical Metallurgy group and are not reported here since they have been extensively published elsewhere (Willies 1990, 1991, 1995). The soundings in the mine have provided important information about the tools utilized in the technology of Bronze Age mining and ore pulverization and indicated that the mine provided some amount of habitation or shelter. It is apparent from the stone tool types found inside that hammerstones were being utilized for battering and pulverizing the ore. But surprisingly, bucking stones, a stone with one flat surface and a hollow in the middle, indicate that grinding also took place. Therefore, the ore was battered, pecked, and enriched inside the mine. The predominance of Early Bronze Age sherds suggests that the third millennium B.C. was the main era of exploitation, but Classical and Byzantine admixture indicates later activity of some type.

Burial Chambers

In the process of mapping and excavating test trenches inside the galleries in 1991, a necropolis/burial chamber was discovered in abandoned mine shafts (Mine 2) and was the target of excavations in 1996 (Willies 1995, Andrews 1994) (Fig. 16). Mortuary traditions as well as data on status, diet, and population can be derived from the analysis of this new evidence. One extensive mortuary chamber contained a number of different burial traditions—pithos burials, stone-built tombs, simple internment, and rock-cut chamber tombs—spanning a date from Late Chalcolithic/Early Bronze I to the early second millennium B.C. A limited sampling from 1992 showed a wide range of ages and grave goods. Although disturbed in antiquity, there was ample indication that intact burials might exist below the rubble accumulation in the chamber. Broken fragments of human skeletal material occurred throughout the chambers. A minimum of 8 individuals were interred in the burial chamber, based on counts and aging information from femurs, mandibles, and a few pelvises. So far this small demographic sample contains children, men, and at least one woman, mirroring a true population composition. One is an infant less than 2 years old. Three are sub-adults less than 18 years old: one 12-15 years old, one from 5-10 years old, and the other approximately 8. Four adults are represented, including 1 female and 1 probable male.[9]

Pottery in association with the burial chamber was chronologically and stylistically parallel with the Göltepe and Kestel sounding assemblages. Pottery such as red and black wares, light clay miniature lug ware, painted Anatolian metallic wares, and imported ceramics such as Syrian bottles and Syrian metallic wares link the mine not only with the processing site,

---

[9] Human skeletal analysis done by Jennifer Jones.

Göltepe, but with neighboring regions as well. However, relative dating through the ceramics suggests that an earlier phase is represented in this chamber, which puts the initial mining of this site into the late fourth millennium B.C. An over-fired greenish Uruk-like sherd found in 1992 also falls into this Late Chalcolithic/EB I horizon (for comparable period and wares see Palmieri 1981, Algaze 1993). A copper spiral akin to examples dating to the end of the third millennium (Goldman 1956: Pl. 432: no. 259), as well as Syrian bottles (Goldman 1956: Pl. 268: no. 617), plain simple ware, and Syrian metallic wares indicate interregional connections with Syria, Mesopotamia, and the Mediterranean coast.

In 1996, the final excavation season at Kestel was completed by a joint University of Chicago/Boğaziçi University team, joined by specialist mining archaeologists from the Peak District Mining Museum in the U.K. The Kestel program aimed at excavating the graves and related features in the Mortuary Chamber, which was first discovered in 1991. This abandoned mine shaft had evidently been reused in antiquity as a graveyard. Mapping continued of surface features which were related to ore processing and open-work mining above the mine on the mountain slope. These areas around the entrance of Kestel Mine 1 and Mine 2 were targeted for excavation to better understand the initial ore extraction methods.

A trench was opened in the eastern end of the abandoned mine shaft, Kestel 2, Mortuary Chamber, and at least three phases of use were identified in the stratified excavation sequence (Yener 1997a and b). The first and lowest phase constituted the extraction of ores, replete with rubble associated with mining. Early Bronze Age pottery fragments were identical to the types found at Göltepe, thus dating the mining in this gallery to the third millennium B.C. There had been substantial domestic use of the underground workings which perhaps were even used for refuge. Inside Mine 2 at least two semi-subterranean pithouse structures constructed of stones were built in the mine shaft after mining had ceased. These two pithouses were similar to the structures excavated at Göltepe and again contained stylistically similar Early Bronze Age ceramics. Finds also included a copper-based pin, a hematite weight, small amounts of antler, and an oven. Postdating the pithouse structures were the inhumations. The furthermost extent of the Mortuary Chamber had a number of disarticulated human bones. Approximately a dozen persons had been buried in pits or extraction cavities. The ceramics found in association with this level indicates an Early Bronze III date for these graves. There had been later disturbance of at least some skeletal remains and probably breaking down of barriers separating inhumation areas from the rest of the mine workings. The human skeletal material had probably been robbed in antiquity or perhaps carnivorous animals scattered the remains around the chamber.

Mine 1 was also later used as a shelter. Earlier excavations in Mine 1 (Willies 1993, 1994), notably Trench 5, indicated use of the mine from the Byzantine period through modern times with no mining rubble associated with these levels. An adjacent, larger chamber had been modified by leveling the floor which had a surface scatter of pottery sherds. More recent use has been by animals leaving a variety of bones and coprolites.

At the Kestel surface, several trenches were put in to investigate the function and dating of the ore-processing features surrounding the entrances of the mine shafts and open-pit mining zones. Trench T10 investigated the surface entrance of the Mortuary Chamber at Kestel Mine 2. While sinking the shaft into Mine 2 to gain entrance into the mine from the surface, a mixed level of fill was found containing pottery, antler, spindle whorls, and bones. When the trench was expanded, this area revealed an oven, suggesting domestic use of the entrance area. Small scrappy walls of stone and several subphases at the entrance of the mine indicated that certain organizational changes had taken place during the Bronze Age. Substantiating earlier crucible finds, refractory crucible fragments, possibly from smelting activities at the surface, suggest that initial smelting occurred near the mine as well as on Göltepe hill. It is possible that the crucibles at Kestel were used to assay the ore for tin content in order to make strategic decisions during mining.

Again at the surface, another trench (T26) investigated the lower open-working area. A large stone mortar was found in situ with a central hollow shaped like a big foot. This was presumably used to crush and grind the ore to render it to a powdery consistency for ultimate smelting purposes. Trench T27 was placed at the original entrance of Kestel Mine 1 where an ore processing station was located. This work station demonstrated how cleverly the angle of the slope may have been used to wash the ore downslope and separate the tin from the iron and quartz by gravity. Ceramics found during the excavation of this trench demonstrated the contemporaneity of the workings to Kestel Mine 1 and Göltepe.

More open workings are located in the broad shallow valley east of the hill and south of the Mine 2 entrance (II to V). Agricultural use has modified the waste heaps, although the working faces are evident in the small western escarpment. Substantial open-work sites are found in the east (VI), north (VII), and west (VIII to XI), upslope as well. Characteristically sub-circular in shape with a working face uphill, a crescentic dump is evident on the down-side. In the hollow, depressed center, a "working area" of broken stone can be seen. Some workings appear to have been cut down to bedrock under alluvial waste, while others were cut into the marble for a few meters depth. Some cut through older underground workings where they were shallow, especially through the small-scale workings northwest of Mine 1 (which may suggest contemporaneity with the large-scale underground workings which do the same). The large extraction areas on the northwest are in an as-yet undetermined strata, probably mainly quartzite, perhaps following a fault structure for ore.

The total volume of open-work type extraction cavities is still in a preliminary estimation stage. Extraction figures for individual open-work sites, neglecting very small ones, range from around 1000 tons to at least 15,000 tons. Technically a much lower-grade ore was extracted from the open working on the surface than from the underground workings, though once enriched deposits were reached, the yields could be high. Perhaps ten times as much ore was excavated at surface quarries than underground and if, as estimated, the ore yielded 10% as high as the galleries, a further 100 tons could be added to the total. Tentatively, total production estimates by the U.K. mining historian/archaeologists suggest a minimum yield for the whole Kestel site of around 200 tons of tin produced over perhaps a thousand years. Working such a low-grade tin site was obviously worth the effort since tin still was a very rare and expensive commodity in the Early Bronze Age. Recovery of very small amounts of gold and use of hematite for pigment is also likely.

The sequence of ore production thus began at Kestel mine and openwork mining areas on the slope. Preliminary ore treatment was mainly at Kestel with final processing and smelting mainly at Göltepe. The termination of mining activity and the production site at the end of the third millennium B.C. suggests the discovery elsewhere of better-grade deposits and the arrival in the regional market of competing, cheaper supplies possibly brought in by the Assyrian traders. It is also possible that the final stages of working were marked by dramatic climatic events which disrupted trade, a matter of much debate lately.

Intensive Surface Survey at Göltepe

While investigations at Kestel mine were continuing, a decision was made to intensively survey Göltepe, the hillside immediately opposite the mine. This decision was based on a reconnaissance on the site in 1988, which had yielded the elusive third millennium B.C. sherds—much searched for but not found in the earlier stages of the regional study. The immediate aim of the surface survey at neighboring Göltepe (Ç13) was to provide information about the lateral spread and density distributions of material evident on the surface. Detailed examination of the morphology of the site showed that cultural strata was situated on a large, battleship-shaped, natural hill with deposition throughout the entire extent of its surface (ca. 60 ha). Determining the horizontal extent of the site was an important goal since a size of 60 hectares is an anomaly in such an agriculturally unfavorable environment. The other aim was horizontally exposing as much of an area as possible to obtain settlement layout and densities, differential quarters with special functions, and to understand the use of space and distributions of artifacts.

The natural hill on which Göltepe is located is geomorphologically distinct from the Kestel tin mine slope, and is mostly a softer shale and greywacke. West of both sites is a third geological subdivision, of diabase/gabbro, the source of the ore-dressing stones. In the immediate environs of the site, the Kuruçay stream flows from the Niğde Massif to the north and spills into the Ecemiş River to the south. Natural springs and agricultural plots surround Göltepe on all sides, while pockets of well-watered, agriculturally richer lands exist 4 km to the southeast near the Ecemiş River and over the Massif in the vast Niğde and Konya plains.

The survey encompassed a 1 km^2 area and differed from the Kestel slope survey in terms of time spent and method. In addition to its location and the third millennium B.C. time span represented by surface finds, the unique morphology of Göltepe was a factor contributing to the selection of the site for survey. Its distinction was its large size, the lack of heavy overburden, and the extraordinary richness of ore-dressing equipment on its surface. The concrete Turkish government mapping elevation post at 1767 m altitude was used as the datum point, A (Fig. 17). Circles were mapped at intervals of 50 meters in the four directions north, south, east, and west from points A to O. A total of 77 large (625 m^2, radius of 14.1 m) and small (100 m^2, radius of 5.45 m) circular sample units were generated in this manner. At first all the circles were measured so that each would have a radius of 14.10 m with an area totaling 625 m^2, or the area of one triangle in a grid at Kestel. This was chosen so that densities and distributions of finds could be compared to the Kestel survey. After 18 circles had been sampled, material and time constraints forced a reduction of the sample size to circles with a radius of 5.45 m radius (100 m^2). The furthest extent of collection to the south was 700 m from datum A. The gentle, undulating spurs at the northern end of Göltepe were covered by eight transect lines diverging at 15° between the east and west axes. Nine axes radiating out from the datum point were thus created, each sample circle numbered consecutively downslope.

The results of the surface survey showed that the summit was badly denuded and outcrops of the flysch/greywacke bedrock at a 20° ubiquitous dip to the south could be seen scattered throughout the hilltop. In most cases these outcroppings could be correlated with greater densities of artifacts, which suggested faster deflation, or stone quarrying by local villagers. With this evidence in mind, subsequent selection of excavation trenches in 1990 avoided these bedrock outcrops as much as possible in the search for depth of deposit, only to realize that the bedrock had been intentionally trenched in antiquity. Subterranean structural units had been cut into the underlying basal clays using the outcroppings as walls, that is, the cultural deposition consisted of the accumulation of strata all cut into underlying geologic sediments.

The entire slope had been plowed up until recent times, and approximately 100 m below the summit on the west slope and 100 m to the south, the remains of the latest phase of the 3rd millennium B.C. occupation lay immediately beneath a 10-25 cm plowzone. What were at first thought to be Byzantine sherds in 1987 and 1988 were later recognized as Early Bronze Age ceramics with eroded surfaces and large, coarse, straw-and-grit-tempered crucibles with a vitrified accretion (Fig. 12: D, H). Recognition of these highland and special function assemblages without local stratified sequences hampered the publication of this pottery. The closest relevant sites are Mersin and Tarsus on the Mediterranean coast 60 km to the south. Once recognized, no pottery of later periods was identified on the surface during the survey, and thus no heavy overlay of later material would hinder the aim of reaching the Early Bronze Age levels.

The link between Kestel mine and Göltepe is demonstrated by their contemporaneity on the basis of pottery and radiocarbon dates, as well as the presence of similar ore and ore-dressing equipment. The particular requirement that ore from tin lodes has to be crushed to liberate the cassiterite before it can be dressed to usable grade makes it possible that Göltepe undertook the final stages of dressing Kestel ore after it had been rough crushed and hand sorted at the mines. This is also indicated by the relative proportions of heavier pounders found at Kestel slope on survey (77%) as opposed to larger proportions of grinding stones (91%) at Göltepe (Hard and Yener 1991). At Kestel, ore would have been broken down using the edges and ends of the stone tools, as indicated by the battering. Further reduction into small pieces and minimal grinding at Kestel is indicated by the presence of bucking stones, or tools with hollows and some grinding surfaces. These two stages would occur when the high-grade ore was selected, while the waste would be discarded. After the ore was crushed and sorted at Kestel, it would have been transported to Göltepe for further crushing and grinding before being smelted in the crucibles. The major components of metallurgy such as fuel and ores with additional complementaries of groundstones are all evident. Final dressing to a concentrate form could well have been a function at Göltepe as water from the streams would have been available during the rainy season.

Göltepe, Tin Smelting Workshops, and Habitation

The archaeological program aimed to recover information pertinent to the techniques of manufacturing metals and the factors that constitute the formation of specialization in a metalliferous zone. By tracing community or workshop patterns through an examination of internal variability and their changes over a relatively short span of time we sought to discover the organizational strategies behind this industry in what appears to be a combined workshop-habitation site. To this end, a total of 1550 m^2 was excavated at Göltepe in 1990, 1991, and 1993 (Fig. 18), while 2500 m^2 were tested through magnetic resistivity, which indicated anomalies where subsurface features existed. In tandem with this, 1 x 1 m test pits and thirty-six stratigraphic profile trenches were executed in a radial configuration around the site to determine the extent and nature of the site. The results of these procedures established that the site was surrounded by a circuit wall. The area of densest population, which was walled in at the summit, measured 5 hectares; less dense, scattered extramural settlement extended to 10 hectares. This is a conservative estimate and it is possible that still more pithouse structures existed between the site and the mining complex. These estimates, of course, do not include the Kestel mine with its 1 kilometer slope area of processing installations, on which evidence of habitation and contemporary pottery were also found. Thus, linked together as an integrated man-mine system, a closer estimate of the total activity zone is probably 60 hectares. The data reported here provides a closer look at the chronological relationship between operations in the mine, at the slope workshop units, and a related industry at Göltepe.

The dating of Göltepe relies upon the stratigraphy of the finds as well as on a series of radiocarbon dates (Table 4). It is important to point out that the Early Bronze Age ceramic assemblage at Göltepe is similar to the finds at Kestel and parallels known sequences at Tarsus, Mersin, and Kültepe, which have comparable chronological spans. The main tradition at Göltepe is a dark burnished ware, which may be a somewhat more refined continuation of similar Chalcolithic ware types (Aksoy 1998, Aksoy and Dupres in prep). Red, brown, and black colors are favored. The vessels are tempered with fine grit and some chaff and are relatively evenly fired. The closest parallels for this pottery come from Early Bronze I and II Tarsus for the plain black burnished (Goldman 1956: 100-101, Fig. 239:

nos 73, 74, Fig. 349: 310) and red burnished types (Goldman 1956: 96, Fig. 241: 92-96).

A second major tradition is a micaceous ware group made with grit temper in colors varying from buff to orange. Fine wares include Anatolian clinky metallic ware (Fig. 12: C, G), and fine slipped wares. The Anatolian metallic wares are made with well-levigated clay, with the buff surfaces usually painted with purplish, brown bands and dripping lines. Quite often there are incised marks on the handles, often on one-handled cups or small jars (Goldman 1956: Fig. 247, called light clay miniature lug ware). It is extremely even and hard fired, giving a metallic, clinky ring. Distributed widely in the Niğde and Konya regions, as well as the central Taurus range, this pottery was first recovered on survey (Mellaart 1954: 191-194, 209, Seton-Williams 1954). Examples have been excavated at the Sarıkaya Palace soundings at Acemhöyük from levels X-VIII (Özten 1989) and these have been compared to ones dated to Early Bronze II and III at Kültepe (T. Özgüç 1986: 38-39: Figs. 3-21) and at Karahöyük level VII (Alp 1968: 304, taf. 10/19). Mellink (1989: 322) describes this pottery as an import to Tarsus from the Taurus mountains and adjoining plateau and suggests connections with the metal sources since some examples were found in the silver mining district of "Bulgar Maden" [now Bolkardağ] (Goldman 1956: 107).

The less common types include fine slipped ware which is well levigated and tempered with fine grit and usually red slipped or painted. Some sherds of plain simple ware and Syrian metallic ware (bottle fragments) were also found, again linking this assemblage to Cilician, Amuq, and north Syrian late-third millennium assemblages (Kühne 1976: 64-65). Orange gritty wares, often large-sized storage vessels, have the same hard-fired characteristics of the Anatolian clinky metallic wares, but the grit and lime particles are usually medium and large sized. The coarse ware group includes cooking wares and baking trays, which are tempered with medium- to large-sized grits and some chaff. The predominant vessel forms are open mouth, shallow bowls, deep bowls, bowls with thickened rims, baking trays, and one-handled "measuring" cups. Closed shapes include beak-spouted pitchers, jugs with a neck flaring from the shoulder, jugs with a neck flaring from the rim, and jugs with a straight neck. Cooking pots, all hole-mouth shapes, are of the dark burnished tradition, mostly buff, tan, and dark in color.

The site was divided into four main sectors, reflecting either different erosional patterns, and thus variegated depths, or differential morphologies, and thus separate quarters. The sectors included: 1) Area A, the gradual step-like slope of the southern slope and summit; 2) Area B, the sharp drop on the western slope; 3), Areas C and D, the gentler descent on the eastern slope; and 4) Area E, the lower western terraces near the stream at the foot of the hill (Fig. 19). Not surprisingly, a preliminary spatial distribution of metallurgical debris inside pithouse structures at Göltepe revealed a special function settlement with a profound association with intensive ore processing and smelting. Larger community patterns, zonal urban quarters, and other aspects of intra-site organization are indeed indicated, not only by the morphology of Göltepe, but also by differences in architectural units in trenches A23, A24, A06, A02, A03, A22, A15 (domestic and specialized), the split-level B05/B06 (public), and E69 (specialized). This variation is further supported by find contexts: concentrations of crucible fragments, molds (Fig. 20), dressing stones, and ores in Areas A and B, the large scale mortars and workshop stalls in A15, A23, and AO2, and the pithos storage jars and domestic utensils in E63, D67, and A06.

Organization of domestic areas, storage facilities, and workshops, and the quantities and variety of goods within these areas will ultimately provide comparative data for the final publication. For the purposes of delineating differences of function within the structures, only a small subset of the excavated units are presented in this chapter. Two neighborhoods of subterranean and semi-subterranean pithouse structures which may have functioned as combined workshop and habitation units in Area A and Area B are taken as case studies. The units are ovoid pithouses which have been cut into the underlying greywacke bedrock with smaller subsidiary bell-shaped pits in association with them. Smaller houses measure 4-6 meters in diameter. Larger units measure 9 x 7 meters and are terraced off the west slope, Area B, much like the layout of the neighboring mountain side village, Celaller. The superstructure of these units is wattle and daub and the great numbers of branch impressions on mud and vitrified structural daub substantiate this. These impressions may enable us to reconstruct the shape or pitch of the roof. The units were plastered repeatedly and up to 25 layers of plaster could be identified. Most of the pithouse structures have a considerable quantity of vitrified clay and roof fall lying immediately above the floors, which seals a number of contexts and provides evidence of burning.

Interestingly, there is a lack of any clearly definable furnaces. Instead, small semi-circular domed hearths were built into the walls and lined with clay and several examples of moveable braziers were found. Unique also are what appear to be geometrically decorated clay panels, which may have adorned the interior spaces of the pithouse structures or provided decorative borders for doors, bins, and hearths. Area E is metallurgically relevant because of the large middens situated on the southwest slope. A midden containing thousands of crucible fragments of the larger variety (20-50 cm) and a great amount of powdered ore was unearthed and 30 kilos of this were taken as samples for analysis. A great many refuse pits yielded debris of a typical metallurgical nature. In a bell-shaped pit underlying the midden, crucibles with smaller diameters were found, suggesting that the size of crucibles may have increased over time.

Architecturally unlike any other prehistoric site in Turkey, the nature of the construction techniques of these subterranean houses demonstrates the long continuity of building by carving out and shaping the local landscape, taking advantage of the natural volcanic topography of the area, a building type that is characteristic of the Cappadocian early Christian churches, monasteries, and settlements best typified at Ürgüp, Göreme, and Kaymaklı immediately to the north in Nevşehir. Some parallels for subterranean structures of this nature were found in earlier periods in neighboring areas such as the Chalcolithic examples in Cyprus and the Beersheva culture in Israel and the late third millennium levels of Arslantepe at Malatya (Frangipane 1992: 184). Given the nature of the intermontane climate, cold and windy at times even in the summer, the subterranean units at Göltepe should perhaps be associated with dwellings suitable for inhospitable mountainous areas and environmentally determined.

Area A and Area B Pithouse Structures

The finds from Göltepe pithouse structures in Areas A and B exhibit a great diversity of metallurgical activities related to tin production. Of special interest are the variety of metallurgical residues, such as lumps of hematite (iron oxide) ore, different grades and colors of ground ore and slag, and metal artifacts. Close to 120 kg of ground and chunks of ore were recorded and large amounts were taken as samples from various loci. A sampling procedure was designed to include all types of primary ore and processing debris materials, as well as crucible fragments, which now total a metric ton.

The find places of crucibles varied from multicelled pithouse structures in Areas A-D to dumps along the circuit wall in Areas B and E. Pulverized ore of a fine powdery consistency, containing from 0.3-1.8% tin, was discovered in measuring cups in sealed deposits on the floors of Area A pithouse structures. Other indications of ore processing and metal production came from the more interesting ore-dressing equipment such as large mortar and pestles (weighing 26.3 kg and 6.5 kg) which were used for crushing the ore (Kaptan 1990b: 28: no. 10). Querns (Kaptan 1990b: 29: no. 11), grinding stones with multiple flat facets (Kaptan 1990b: 29L: no. 12) (Fig. 3: A, F, G), polishing stones (Fig. 3: C), and large groundstone axes weighing 2.9 kg (Kaptan 1990: 30: no. 16) (Fig. 12: M), as well as bucking stones with multiple hollows (Kaptan 1990b: Figs. 13-15) also indicate an important commitment to industry. A number of sandstone molds with bar-shaped beds carved on several surfaces suggest that tin metal was being produced and poured into ingot form before being transshipped to locations for bronze alloying. Multifaceted molds such as these are typical of the late third-early second millennium B.C. in a number of sites in Anatolia (Goldman 1956: 304, Braidwood and Braidwood 1960: Fig. 350: 1). Since the necropolis of the site has not been located, the only metals found were small scale pins, awls, rings, and other fragments. Analyzed with AAS by Özbal, all contained between 4.75-12.3% tin, demonstrating the high tin content of the metal unearthed at the site (Table 5). Interestingly elevated levels of gold were also observed, suggesting the possibility that Kestel was the source of the tin used, since gold is a component of this deposit as well.

Only a small portion of the household data is presented here since a number of dissertations and research projects are in progress. This will provide a preliminary model from which to infer the relationships between particular pithouses and/or neighborhoods vis-à-vis the intensity of metallurgical versus domestic activities. In any case, an important prerequisite for making comparisons with other metal production areas is the establishment of reliable information about floor assemblages and production organization. What seems to be a concentration of pithouse structures related to a possible workstation sector was exposed in the southern end of the summit, Area A. These nine pithouse structures appear to have been used for both habitation and work since workshop features could be inferred from the contents. Sub-rectangular in shape, the pithouse units were cut into the bedrock with an approximate total area of 6 m^2. In some instances a dry laid stone wall served to buttress the crumbling bedrock. Three pithouse structures, Pithouses 6, 22 and 15, were of special interest since they appeared to contain workshops, storage areas, and domestic quarters and were perhaps part of a multiroomed complex (Fig. 21).

Pithouse 6 is roughly oval in shape and measures approximately 2 x 2.8 m in area and perhaps only 80 centimeters below ground level at its deepest point. A vast quantity of mud daub chunks, many preserving the impressions of wooden poles, were recovered, providing further support to the belief that these structures possessed wattle and daub superstructures. Although no post holes were observed, some pithouses had a flagstone in the center or along the side of the room, suggesting that wooden posts would have been placed on these as a column base. In the northeastern corner of the structure there was a shallow pit that contained a cache of 9 groundstone tools. Collapsed over this pit was a large, clay panel with geometric relief decorations on one side (Plate 8). Conspicuous by its absence was a hearth or furnace. The burning and consequent collapse of the roof preserved a rather substantial floor assemblage including numerous groundstone tools, a large fragment of a crucible, complete ceramic vessels (including a large pithos and an Anatolian metallic ware pitcher), several large, baked-clay blocks of an unknown function, and perhaps most interesting, a vessel with a flaring rim containing approximately 9 kilos of ground ore material (Plate 9). Other metallurgical materials included groundstone ore crushers, mortars, bucking stones, and kilos of ground ore and chunks of ore. These appear to represent the tool kit of an individual or group of individuals engaged in the final preparation of the cassiterite ore prior to smelting. Most of the floor artifacts were recovered from the edges of the structure, including the nine groundstone tools found in Pit 7, leaving a small area in the center of the structure that may have served as a work area.

Pithouse 15 (Fig. 21), located adjacent to Pithouse 6, was perhaps part of a split-level multiroomed complex. The floor is approximately 1.5 m below the edge of the cut bedrock. It was also clearly destroyed by fire, sealing a considerable floor assemblage. The structure itself is in the shape of an elongated oval, measuring approximately 3.5 x 2 m. Two post holes were observed interior to the southern margin of the structure. Excavations within this structure also provided ample evidence of the nature of the superstructure in the form of numerous pieces of structural daub, as well as several panel-like chunks of daub which may have been part of the roof. Well-preserved pole impressions were observed on these panels, including impressions of the material used to strap the poles together, leading us to conclude that the roof in this case was flat.

The floor assemblage reflects a tool kit associated with the final processing of Kestel tin-rich ore. In addition to several groundstone tools (including bucking stones, hand-held grinding stones, and a part of a slab mortar), a complete conical crucible was recovered near the center of the room (Plate 10). The crucible was found lying on its side, with several flat pieces of sandstone partially covering its opening. Several more flat pieces of sandstone were found resting on the floor below the crucible. The large, conical crucible had not yet been used in smelting since the ubiquitous gray tin-rich layer was not observed on its interior surface, although stone covers were found in situ. A hand-lens revealed globules of vitrified material adhering to the inside surface of the stone covers. This supports the hypothesis that the flat stones were placed over the mouth of the crucible as a means of retaining heat during the smelting process. It is possible that this vessel was used to roast the ore or that the roof collapse occurred prior to the final smelting of the ground ore. No hearth or furnace was found in this pithouse structure. However, a portable oven or brazier was recovered resting on the floor near the center of the structure.

Structures built on the sharply sloping western slope in Area B (Fig. 22) presented evidence of different architecture and greater depth of archaeological deposit. Larger structures were found on this side of the site. The largest, approximately 24 m², had a wall on the west side but was cut into the bedrock on the north and south sides. No wall was found on the western, downslope side, and thus the method of roofing and the exact size of the unit is still unknown. A marked difference was observed in the carving of the basal clay bedrock. Three terraces had been cut into the bedrock on the slope for leveling and structures were constructed into each of the terraces. One terrace was trenched up to four meters and walls were erected in front and parallel to these cuts in a substantial slope structure (House B05 and B06). This mode of construction is still in evidence in Celaller village, which gives the appearance of extensive slope trenching to emplace houses similar to a staircase, each roof serving as the front entrance of the house above.

Structure B05 contained a plastered feature with three compartments, which was built with small stones and reused clay blocks with geometric designs and was found in the north side of the room. Two pots were dug into the floor which was strewn with layers of ground ore. These ore deposits were layered with alternating dark and light ground powder and may have resulted from the alternating heavy/light fraction of a vanning procedure. Ten kilos of ground ore were taken for analysis. Groundstone tools and ceramics were found on the floor and in the fill above the floor. Another room downstairs or a perhaps separate independent structure (B06) was located 1 m below B05 to the south, making this a possible second example of the split-level architectural style mentioned in conjunction with Area A pithouses above. A geometrically decorated hearth slightly off center and a pyrotechnological feature in the northeast corner were found in situ (Plate 11). Finds from the floor of this structure included groundstone mortars, grinders with ore still on the underside of stones, kilos of powdered ore, crucible fragments, a lead ingot weighing 170 g, and a silver, coiled-torque necklace (Fig. 23) made of an unusual alloy containing tin, copper, and zinc in high levels (Yener, Jett, and Adriaens 1995: 72).[10]

Ringing the subterranean workshop/habitation dwellings on the summit was a well-built north-south, perhaps circuit, wall on the west slope, which was preserved in some places to ten courses and a height of over a meter. The plan of the circuit wall resembles the zig-zag pattern of Early Bronze II fortifications at Tarsus (Goldman 1956: Plan 6). Constructed of large, irregularly shaped stones, parts of the wall were built with reused saddle

---

[10] AAS analyses by Özbal determined 91.2% Ag, 2.18% Cu, 1.16% Zn, 1.11% Sn, 0.41% Fe, 0.25% Sb, 0.18% Bi, 0.02% Ni, and 0.01% Pb. This rather high tin content in a silver necklace points to the availability of tin in this region.

querns and large groundstone mortars with hollowed surfaces. Since the western slope did indeed yield the remains of a more substantial public building complex surrounded by domestic and/or industrial quarters on the spurs and terraces, one of the most important questions to be answered is the apparently distinct, separate role played by the workshops along the slopes of the Kestel mine and the linkage between them and the workshops on Göltepe. Possible interpretations for this distinction are: a) a separate year-round settlement at Göltepe with industrial sectors for fine dressing and casting of metal; b) a separate ceremonial sector located in Area B; c) a seasonal settlement and heavy-crushing workshops at Kestel; d) a commercial sector or trading harbor at the larger site on the lower slopes near the mountain passes.

Taking the organization of the industrial complex as a whole and integrating the activities of Kestel and Göltepe, as presently understood, Kestel mine was originally in operation around 3,000 B.C., probably as an open pit mine. In the early EBI/II, the mine was expanded into shaft and gallery systems, work stations were set up outside the mine entrance, and Göltepe was settled. Göltepe grew into a (possibly) substantial walled town and tin processing workshop site near the middle of the third millennium or Early Bronze II, at the same time that bronze was becoming more widely used in Anatolia and its use was becoming more widespread in Syria and Mesopotamia. The settlement and the mine attained their largest extent during this period and the early part of the EB III period. At this time, these sites were one of a number of centers that lined the strategic passes through the Taurus mountains and that may have controlled both the resources, the production, and the intermontane traffic during the pre-Akkadian period. Kestel mine may have continued its existence into the early second millennium B.C., as indicated by ceramics.

It is apparent that the entire site of Göltepe was occupied during the same period (with at least two distinct occupational phases everywhere)—the early/mid-third and mid/late-third millennium B.C. There is no definite evidence for earlier or later occupation or disturbance from either surface survey or from excavated trenches. The area enclosed by the circuit wall suggests that all sectors were part of a single settlement rather than complementary and alternating smaller settlements. The remains of a larger public building with domestic and/or industrial quarters on the spurs and terraces of B05 suggest that tin processing occurred both in the larger units as well as within the smaller pithouse structures along the southern summit in Area A.

Clearly tin processing was the special function of Göltepe and, to judge from the quantities, was produced on a large scale. The next chapter will present the instrumental analyses of the various crucibles, ore and slag powders, and other metallurgical residues. These have been used as a frame of reference in an attempt to recreate the production technologies through several smelting experiments. This in turn will hopefully provide important clues to the organization of the production industry. The habitation/workshop site of Göltepe was a first-tier, industrial operation focused specifically on processing tin ore. The ore was then transported elsewhere, presumably to a second-tier, lowland, urban workshop as yet not identified,[11] for the next stage of processing which was alloying and casting it into a diversity of artifacts. Although the production model at Göltepe and Kestel can be typified as a cottage industry, nevertheless, these mining/processing sites have yielded substantial evidence that the output was quite large.

---

[11] The results of lead isotope analyses conducted on the silver helmet of the Amuq phase G male figurine and a dagger from phase F indicate the central Taurus as the source of the lead. Equally intriguing are several silver fragments from Middle Bronze Age Acemhöyük located in central Anatolia also stem from the Taurus. This suggests that the lowland workshops were located in a number of different directions *see* Yener *et al.* 1991, Sayre *et al.* 1992.

CHAPTER FOUR

THE PRODUCTION OF TIN

The Smelting Process

When excavations commenced at Göltepe in 1990 for the purpose of investigating the village of miners, no visible signs of smelting, such as mounds of slag, which are generally produced when copper, silver and iron is smelted, were evident. Thus, assuming the site only functioned as a locus for habitation, and that the smelting site was located elsewhere at an as-yet unfound site, the initial excavation design targeted domestic activities related to subsistence. Moreover, it was assumed that more commonly known fist-sized slag would be inaccessible under thousands of tons of Ottoman silver smelting slag in the village of Çamardı four kilometers away (Yener *et al.* 1991). To add to the frustration, recognizable vitrified and slagged furnace fragments were not found on the site surface survey either. As the low tin assays of Kestel mine increased skepticism about the actual metal produced in the increasingly negative and boisterous literature and even e-mail, it was strikingly obvious that tin smelting correlates in historical contexts elsewhere had to be researched first before proceeding further.

Consider the obstacles hampering initial attempts to understand the nature of tin technology in its formative periods. First, all evidence of tin smelting in archaeological research to date and even in the historical documents available was on furnace-smelting technologies. This type of bulk smelting generally produces residues such as recognizable lumps of black glassy tin slag, not unlike obsidian. Even a search through the massive resources available to the Smithsonian Institution to find a published photograph of tin slag produced only a few examples, and most of them from 19th century literature. The solution was a search by the author in Cornwall itself (Earl 1985, 1991, 1994), where tin slag and tin smelting sites provided index fossils and comparata to continue the search in the Kestel area. But even in Cornwall, information about prehistoric tin smelting was limited and surprising gaps existed about the nature of prehistoric tin smelting (*see* Penhallurick 1987). Smelting achieved in crucibles had not been considered to be part of the metallurgical repertoire of tin production, and even analyses of tin slag were rare in the literature (Tylecote, Photos, and Earl 1989). Moreover, the fact of crucible smelting (not melting) was only just gaining acceptance in the scholarly community, even with better researched copper-based metallurgy (Tylecote 1974, 1987, Pigott *et al.* 1988, Zwicker 1989, Zwicker *et al.* 1985).

Thus the Göltepe finds, such as the diversity of metallurgical residues, selective lumps of ore, different grades and colors of ground ore and ground slag materials, and a whole spectrum of different crucibles and crucible fragments, were hitherto unknowns. It was obvious that these metallurgical residues represented important facets of the tin production operation, but the production procedures and processes as they changed through time needed to be systematized. As contradictory as this may sound, the overabundance of metallurgical analyses widened the unknowns, resulting in the realization that the nature of prehistoric tin smelting technology and its resulting by-products clearly would have to be redefined. For example, the importance or magnitude of ore preparation and grinding had not been recognized before, even though 50,000 groundstone tools were found on the surface of the site. The need for a powdery consistency as a prerequisite for metal preparation became apparent when 5,000 lithics with grinding surfaces continued to be unearthed from the excavated contexts. That these were vital components of a tool kit for ore dressing and separation of smelted metal from slag became clear when experimental smelting replicated the method (see below) and entrapped cassiterite was found by SEM in the grinding surface of the stone tools. By means of a many-pronged analytical program the *châine d'opératoire* for producing tin was revealed.

Ore Materials from Göltepe

Starting with particular find categories, the analytical program was aimed at ultimately expanding inferences about the organizational strategies of the tin industry. In order to conduct replication experiments the metallurgically relevant materials were first subjected to instrumental analysis. One of the categories of finds was ore lumps, nicknamed nodules, which had been recovered in considerable amounts; these resemble the tin-rich hematite ore at Kestel and in fact actually turned out to be exactly that (Bromley 1992). Analyses of these nodules yielded hematite as the base constituent with an average tin content of 2080 ppm (with a range from 0-14,300 ppm), nearly three times the average remaining today at Kestel mine (Table 6). Analysis of one hematite nodule sample revealed that it contained 1.5% tin, suggesting that at least a 2% or higher ore was mined originally at Kestel, a very good grade of ore (Earl 1994, Earl and

Özbal 1996, Yener *et al.* in press).[1] This demonstrated that only high-tin containing hematite was selectively transported from Kestel mine to Göltepe for processing (grinding) and smelting purposes.

Another important category given extensive analyses was the various powdered materials found in floor assemblages, cups and large storage vessels, and middens (Earl and Özbal 1996, Özbal 1993). Sand-sized ruby-red cassiterite was easily identifiable in samples vanned from powdered materials. The colors of the powdered materials ranged from purple/burgundy, pink, black to beige. Cassiterite grains are easily separated by such manual means as a pan or a vanning shovel and very little water. Due to its high specific gravity, particles of tin separate out from the less dense magnetite, hematite, and quartz. The reddish-colored cassiterite forms a high density "crown" located at the central curvature of the vanning shovel near the rim. When the powdered materials were vanned (concentrated) and the nodules were crushed and vanned, cassiterite again appeared as a distinct, reddish-colored head. These ground ore and/or slag materials contained tin of different grades (0.3-1.8%) and have been identified as unprocessed powdered ore material, tailings from an ore concentration process, and remnants of pyrometallurgical processes (Table 7). When enriched by vanning, the powders showed a tin enrichment by a factor of 5.3 to 11.8; in sample 3842 for example, enrichment jumped tin content from 1.4% to 7.42%.

The powders were further characterized and classified using x-ray fluorescence (XRF), electron probe X-ray microanalysis (EPMA), and X-ray photoelectron spectroscopy (XPS) (Adriaens *et al.* 1999). Nine powdered materials from middens, floors, fill, and inside ceramic cups were analyzed (Table 8a). Tin contents ranged from 0.2 and 2.9% (Table 8b) with relative abundances shown in Fig. 26. Next, analyses were conducted on the tin-containing particles from each powder sample to determine the form in which they occur (Table 9). Figure 27 shows the abundance of each particle group and demonstrates that samples 1, 7, 6, and 3 contain predominantly SnO_2 particles, while the other five samples contain about 50% SnO_2 particles, the rest being tin silicates and Fe-Sn-rich particles. XPS determined the presence of metallic tin in samples 2, 4, and 9 (Table 10) (Plate 18) suggesting that heat had been applied and that some of the powdered materials found at Göltepe were ground slag with the tin metal already removed. Other powders were magnetic, corroborating heat treatment to at least 600° C.

---

[1] Earl notes that any deposit of tin with greater than 5% average of tin is rare, 3% is considered rich ore, and 0.2% has been and can be mined from stream deposits with simple vanning techniques.

Perhaps the best indication of processing aims at Kestel and Göltepe was the undeniable increase of tin content in a flow pattern starting from vein samples taken in the mine, to samples from the hematite ore nodules found at Göltepe, and finally to samples of the multicolored ground and pulverized ore found stored in vessels and floors of pithouse structures. Tin-rich hematite was being enriched between the mine and the smelting crucible. None of the other elements analyzed showed this patterned increase. The lowest tin content in a ground material was analyzed in samples taken from the middens (garbage). Clearly tin had been extracted after heating, and the dross disposed of in dumps or garbage pits. The presence of a metallic tin phase in these powders could easily be concentrated by simple metallurgical processes.

A more cogent mistaken notion is that smelting cassiterite with such high amounts of iron as an impurity is impossible. Analysis of the metallurgical debris from Göltepe has given substantial evidence that the ore was heated during processing because it is attracted to a magnet, and therefore contains magnetite. Magnetite does not occur in the mine suggesting that the magnetism was artificially induced by heating the hematite (Fe_2O_3) ore into magnetite (Fe_3O_4). After grinding the ore, this transformation is easily achieved at 550° C by roasting it inside a bowl, perhaps one of the larger-format ceramic coursewares. The next stage would entail swirling the preheated powdered ore around in water, causing the magnetic iron to clump together and leave the cassiterite as a residue, a simple magnetic separation which would discard much of the iron. It is also possible to alloy high iron tin with copper to make tin bronze. Ellingham diagrams indicate that it is difficult to alloy copper with tin that has high iron impurities while at equilibrium conditions. Nevertheless, several successful experiments produced bronze with varying low iron traces, which suggests that it is possible, but in conditions that were obviously not at equilibrium (Yener *et al.* in press, Earl and Özbal 1996).

Another paradox for those not familiar with cassiterite smelting products (Muhly *et al.* 1991) or with the paucity of debris resulting from crucible smelting instead of furnace technology (Sharpe and Mittwede 1994) is that these processes do not result in great quantities of slag or other metallurgical debris (Earl 1985, 1991, Earl and Özbal 1996). While it is recognized that cassiterite alone will smelt directly in a crucible, it is less known that such a process requires reduction by carbon-rich gases and would generate little slag. The exact same situation exists in Fenan where it is only in the Early Bronze II/III period, when self-fluxing copper silicates which occur with magnesium oxides were used, that larger slag deposits are found (A. Hauptmann 1995).

The resulting tin metal prills (globules) encased in glassy slag were released by grinding with a lithic tool. The slag was thus in a powdery consistency and virtually invisible. Thus although ordinarily one would expect to see discernible features of industrial production such as slag heaps, especially in a supposed specialized metal production function site, the processes of smelting cassiterite in a crucible would produce hardly any slag. Furthermore, the minute quantities of vitrified materials that were produced in the crucibles would be ground down to release metal prills, thus turning the products to powder as well. Horizons of finely ground ore and vitrified materials can be seen in trench profiles all over Göltepe, and are found strewn on floors of houses, inside cups, stored in vessels, and discarded in midden deposits. There are, indeed, several hundred thousand tons of slag located 6 km away in the nearest town, Çamardı, which has erroneously been used as evidence that silver smelting was the main operation at Kestel (Sharpe and Mittwede 1994). Demonstrating how polymetallic the region is, those slag mounds are from the Bereketli Maden silver mining operations, which is also located at the Niğde Massif and have not been securely dated, although the upper deposits are from the Ottoman period. In the future it would be exceedingly interesting to test the notion of special function sites oriented toward the production of silver and gold as well as copper-based products.

Analysis of the Earthenware Crucible/Bowl Furnaces

The most convincing evidence of tin production are the thousands of crucible fragments with tin-rich slag accretion and pyrotechnological features. Over a ton of vitrified earthenware ceramic fragments with bloated, sometimes vitrified inner surfaces, rich in tin, were a cogent reason for discounting the skepticism of tin production here. However, during the 1990 excavation season, before analyses revealed the tin-rich interiors, the function of the coarse ceramic fragments was unknown. With large tin slag chunks obtained from a Medieval tin smelting site in Crift Farm, Cornwall as comparisons, the search for tin slag by the survey teams in the vicinity of Kestel continued seemingly without success in the early stages of the research. The breakthrough came when visual examination of the ceramic interior surfaces with a hand-held lens revealed the elusive nature of the tin slag at Göltepe—the glassy slag was in small particles. Instead of the expected large-scale furnace products the inner surface of the coarse ceramic bowls contained millimeter-sized vitrified material.

These materials were subjected to intensive examinations by a number of laboratories. Initially only rim sherds were taken to be examined to enable reconstruction of the vessel shape as well. Once these ceramics were recognized as being crucible/bowl furnaces, bases and body sherd samples broadened the sample size for instrumental analysis. Subsequent analytical programs targeted greater varieties in crucible morphology and find places (Adriaens, Yener, and Adams 1999). Thin section, atomic absorption spectography (AAS), SIMS, and microprobe were used to coax out new data. Twenty-four crucible rim fragments were initially analyzed at the Conservation Analytical Laboratory. The following descriptions draw from several articles where greater details can be found (*see* Vandiver *et al.* 1992, Yener and Vandiver 1993a and b, Vandiver *et al.* 1993).

Constructed from a coarse straw-and grit-tempered ware, the crucibles have vitrified inner surfaces containing between 30-90% tin content. During the 1993 season, crucibles were unearthed in a greater diversity of sizes, shapes, and wall thicknesses. The dimensions are variable, but the thickness averaged 1.2 cm, while the preserved height averaged 6.1 cm; the average diameters were estimated at 20-50 cm, and the height ranged from 12 to 40 cm at the rims, some as small as 6 cm. Thinner fragments, averaging 0.9 cm, were also found, leading to the conjecture that at least two functional types of crucibles were made and used (Fig. 24a-d).

The crucible fragments were built by adding slabs often with a strip added to form the rims. Examples found in 1993 indicated that crucibles were also reused and that some examples had been refettered. All shared such distinguishable features as a reduced, hard, and probably high-fired gray inner surface and a much softer, lower-fired, and red or oxidized exterior surface. These characteristics differentiated these fragments from coarse ware cooking pots which are usually harder and reduced on the exterior. Rare examples did not have gray interiors and were either unused crucibles or may have been used to roast the ore in preparation for separation. Impressions of burnt chaff that had been added to the clay were visible by microscope. The function of chaff temper was well understood by the crucible makers since some crucible fragments did not have chaff temper in their inner layers. Closed pores provide good insulation and thermal properties, whereas sand-tempered ceramics with low porosity give more stable, slag-resistant refractories. The Göltepe crucibles/bowl furnaces combined the features of a sand-tempered inner layer and a chaff-tempered outer layer into one product. The smoothed, finely laid inner clay surface also prevented metallic prills from escaping into the fabric of the crucible, a point which became vividly apparent during the smelting experiments.

Functional or chronological differences may underlie the variations among crucible fragments. Some may have functioned as a lid or superstructure, although the 1993 excavation uncovered covers made from flat slabs of stone in Pithouse 15. In other examples where the ceramic surface is over fired, reduced, and bloated, a function as a crucible is more likely. Slag is rarely present in quantities greater than a cubic millimeter, which is understandable in cassiterite crucible smelting. In fact even at Cornwall, in the entire span of tin smelting with furnaces, the greatest volume of slag left today is negligible (Earl 1991). No microscopic evidence of grinding or chipping of slag from the surface was found. Intentional breaking may be indicated by the presence of so many thick fragments of nearly the same size (2.5 to 4 cm thick; about 6 cm maximum diameter). It is also highly possible that different vessel sizes are indicative of the different stages within the smelting process. That is, in a repetitive smelting procedure, crucible size would diminish as the product became more refined, the last stage being a small-scale crucible to melt the metal. An alternative suggestion is that the diminishing grade of ore in Kestel mine warranted the smelting of larger and larger amounts of ore, using larger and larger crucibles. Variations due to chronological factors became immediately apparent when a cache of discarded crucible fragments was recovered from one of the garbage pits in Area E. A midden containing thousands of crucible fragments of the larger variety (20-50 cm) and a great amount of powdered material (30 kilos were taken for analysis) were unearthed (Fig. 25). In a pit underlying the midden, crucibles with smaller diameters (12-15 cm) were found, suggesting that the size of crucibles increased over time.

Variations in firing temperature were evident in the microscopic characteristics of the crucibles. A polished cross-section of the bloated, blackened interior layer was viewed with scanning electron microscopy (SEM) and showed the exterior surface (Yener and Vandiver 1993a: Fig. 10A: upper) and more friable, lower-fired, brown ceramic layer (Yener and Vandiver 1993a: Fig. 10A: lower). A section of the higher-fired surface layer had fused particles and rounded pores (Yener and Vandiver 1993a: Fig. 11). The pores gave a glassy appearance and were produced by bloating during firing at relatively high temperature. By contrast, the low-fired ceramic exterior (Yener and Vandiver 1993a: Fig. 12) showed pores with irregular interiors and fine, clay particles between glass.

Nondestructive X-Ray Fluorescence (XRF) provided elemental analysis of the interior and exterior of each crucible fragment. Elemental tin was detected on the interior surface of 21 of the Smithsonian crucibles but not on any of the exterior surfaces. Iron and calcium were the other major elements present, potassium, titanium, manganese, strontium, and rubidium were usually present, and arsenic was often present in minor amounts (approximately hundredths of a percent). Copper was found in only one analysis (crucible no. 24), a crucible with an atypical texture and composition. Arsenic oxide was found on half of the crucibles. In one case, the atypical crucible no. 24, a relatively high concentration of about 2-

5% As was found in certain areas. The other samples contained only parts per hundred arsenic. On three examples arsenic was found on the outsides but not on the insides of the crucibles; on one, it was found both on the outside and inside. On four it was found on the interiors only. Variable amounts of arsenic can be explained as arsenic that was deposited during the vapor phase. It is possible that arsenic was part of the composition of the original ore (now gone) in Kestel or that it was intentionally added to lower the melting point.

Separate analyses using SEM and EDS (Tracor Northern Energy Dispersive X-Ray system 1700) corroborated the presence of tin on the inner surface of the crucibles as well. The interior surfaces were compared with the exteriors and with the soil in which they were buried. Elemental identification of fine particles by EDS showed tin and calcium present as the major elements with silicon, aluminum, iron, and titanium present in minor concentrations (Yener and Vandiver 1993a: Fig. 14). No tin-containing particles were found on the exteriors of the crucibles or in the soil. The firing temperature of the crucibles was estimated empirically by refiring five crucible fragments to 700, 800, 900, 1000, and 1100° C, followed by microscopic comparison of the variation. At 1050° C the fragments changed color to a glassy, reddish brown found in only a few of the crucibles which did not have a gray interior surface. At 1100° C bloating occured, producing pores much larger than any of those found in the crucibles. Thus 1000° C was estimated as an upper limit for the original firing. The degree of rounding of the particles and pores on the crucible exteriors showed greatest similarity to firing temperatures of 700 to 800° C. The crucibles were possibly set in the ground during the smelting operation in order to maintain lower temperatures. This is inferred from the oxidized surfaces generally found on the exteriors. Furthermore, the crucibles were probably not prefired and then used for tin smelting, but rather, were fired for the first time with the tin ore charge in place (Yener and Vandiver 1993a).

Recent atomic absorption analyses of the crucible fragments support the earlier Smithsonian results with vitrified examples containing up to a 4% tin content. Four of the new samples of crucible fragments tested yielded tin content above 1% (1.009%, 2.09%, 2.21%, and 3.65%), a five-fold increase relative to the powders. This is also verified by a series of analyses using microprobe and Secondary Ion Mass Spectrometer (SIMS) at the University of Chicago and Antwerp (Adriaens 1996, Adriaens, Yener, and Adams 1999, Adriaens *et al.* 1997). The results showed that the fabric of the crucibles consisted mainly of aluminosilicates with fragments of quartz and iron oxides. An accretion layer of calcium carbonate visible on most materials excavated at the site was due to the burial of the material in limestone-rich soil and all of the ceramics at the site had this accretion. Below the calcium carbonate layer, a layer of silicate material with 2-3% tin oxides is apparent; it is represented by a bright band of several micrometer thickness in the backscattered electron image (Plate 12). Small inclusions with up to 40% tin oxides are from a different silicate phase. SIMS was used for line scans across the cross section of the crucible fragment. The bombarding ion beam was moved in distinct 10 micrometer steps across the sample, while acquiring compositional data. A tin peak is clearly present at the interface of the ceramic material and the calcium carbonate layer and therefore at the inner surface of the crucible fragment. The large crucible fragment from Pithouse 15 mentioned above had not been fired, based on visual observation. In order to examine whether tin could also be found in unfired crucible surfaces, analyses were conducted on these as well. This would rule out the possibility that the presence of tin in the crucibles was part of the crucible production process. Indeed, even though tin-rich ground ore was strewn all over the floor, SEM-WDS and SIMS analyses could not demonstrate the presence of tin at all in the unfired crucibles (Adriaens, Yener, and Adams 1999). The house had burned down prior to the firing of the crucible smelt.

One ceramic sample still had remnants of a shiny, glassy, green accretion (a few square cm). SIMS analyses of the vitreous sample indicated it was mainly composed of a silicate matrix containing alkali elements, iron, tin, aluminum, manganese, and titanium (Adriaens *et al.* 1997). Two types of grains are apparent in the matrix in the backscattered electron microprobe image (Plate 13). The equiaxed quadrangular grains are iron and tin oxides, with an average size of roughly 10 mm^2. Longitudinal-shaped grains are composed of tin oxides. These crystals are 0.5-2 mm wide and can be up to 50 mm long. They are similar, but larger than the SnO_2 crystals observed earlier in sub-mm droplets of accretion on crucibles analyzed at the Smithsonian Institution. The glassy accretion is composed of a mixture of silicates, oxides, and metals and, therefore, resembles a typical metallic tin slag (*see* Bachmann 1982). Medieval tin slag from Crift Farm in Cornwall, U.K. was analyzed with SIMS, SEM-WDS as a comparison to the Göltepe crucible accretion (Adriaens 1996). Similar silicates, longitudinal grains, and a similar variety of oxides were present. Some dissimilarities were observed in the different gange materials and a metallic tin prill was observed, which had not been present in the crucibles. This is not surprising given the advanced Medieval smelting techniques at Crift Farm attained using a furnace.

The crucible production model is therefore the following: to smelt cassiterite to metallic tin a temperature of 950° C at a partial pressure of oxygen of 10(-14) atmospheres is needed. To smelt SnO to tin metal involves a much higher temperature range of 1250-1540° C, but a lesser reduction of only 10(-6) to 10(-12) atmosphere. A partial pressure of 10(-4) can be maintained in a smoky hearth or updraft kiln. Copper smelting requires 10(-6) atm. The low-temperature processing parameters suggested by the Göltepe evidence implies that only the first process is possible. Special means were needed to use a low temperature for smelting and achieving a highly reducing atmosphere. Indeed, Göltepe yielded a partially covered crucible packed with a reducing fuel and crucibles which contained vegetal fiber which helped insulate and maintain the atmosphere and temperature. Reduction was achieved with a temperature between 800 and 1000° C (interior temperatures 700 to 800° C for exterior temperatures) for a relatively short duration of a few hours at most, during which time the raw materials sintered and did not entirely melt into a glass. Very small cassiterite crystals, on the order of a few microns, were precipitated (Yener and Vandiver 1993a).

The production model which was suggested by analyses of the crucibles includes a labor intensive, multistep, low-temperature process carried out between 800° and 1000° C. Processing involved intentionally producing tin metal by reduction firing of tin oxide in crucibles, with repeated grinding, washing, panning, and resmelting. The raw materials being processed in the crucibles consisted of tin oxide (cassiterite) with no copper ores present, along with calcium carbonate, iron oxide, and charcoal as the reduction agent. The ore was dressed by grinding and was separated probably during a washing stage such as vanning using density differences. The next step entailed grinding the slag and the material trapped in the surface of the crucible, separating these. There was at least one more smelting step, if not several, in order to increase the size of the tin prills. The high viscosity of the slag in which the growth of tin grains occurred suggested that the smelting process was not very efficient. Increasing the agglomeration of these grains was probably the limiting factor in achieving high yields of tin. The final step in the process is predicted to be the agglomeration of the tin particles and their separation from the slag. This was achieved during a final firing at a low temperature of only 273-300° C at which point the tin metal would have been "sweated" out of the finely crushed slag. It is suggested that these production parameters are profoundly associated with the low-grade tin ores found at Kestel and Göltepe, and not with the formative period when assuredly higher grade, possibly alluvial, deposits which are no longer detectable existed.[2] The fact that the industry lasted for such a long time, and that a labor energy input, albeit difficult to understand in today's standards, was sustained to produce

---

[2] The prior existence of alluvial deposits is based on geological surveys. For details see articles by Willies 1990, 1991, 1992, 1993, 1994, 1995, Earl and Özbal 1996.

tin should be ample indication of how valuable the alloying material was in the Early Bronze Age. In modern times only gold is equivalent to this.

Smelting Experiments

After analyzing a large quantity of the metallurgical residues and by-products of the smelting operations, the stage was set to experimentally recreate the smelting technology using the low-grade tin ore as the charge. Several replication experiments were conducted in conjunction with analytical programs to test the feasibility of the production model, the physical conditions required, and the expected end products. Tin metal was successfully smelted in 1992, 1993, and 1994, utilizing ground materials found in Early Bronze Age contexts at Göltepe. Great care was taken to use the archaeological charge, that is, the material utilized to smelt tin metal by the craftsmen in antiquity. Powdered materials found in one-handled cups and vessels as well as samples found deposited on the floor of the pithouse structures were selected for the experiments. These low-grade ores rather than richer, commercial cassiterite samples were chosen to approximate conditions in place at the final phase of production at the sites, instead of duplicating parameters of the posited, richer alluvial cassiterites. Since very little information about crucible smelting of tin existed prior to this investigation, the experiments soon provided hypothetical production stages and identified the expected archaeological data associated with each stage (Brooks and Yellen 1987, Kramer 1982). Experimental archaeology seeks to define direct relationships between human behavior and material culture, and the caveats were carefully noted (Binford 1987, Hodder 1987, Kent 1990, Seymour and Schiffer 1987).

To this end, a model was constructed for tin production steps at Göltepe based on data generated from the 1990-1993 excavations and laboratory analyses of production debris. A total of four experiments were conducted by tin specialist Bryan Earl from Cornwall, one in Cornwall, two in Turkey at Celaller village, and a fourth in Chicago at the courtyard of the Oriental Institute at the University of Chicago. These and other products were subsequently analyzed by atomic absorption spectography.[3] A video camera documented these replication experiments. Each phase of the production process from mining to finished product was identified, its elements defined, and archaeological and mineralogical implications investigated.

---

[3] AAS by Hadi Özbal of Boğaziçi University in Istanbul. I also thank Judith Todd and Gary Laughlin of the Illinois Institute of Technology who contributed to the understanding of this production.

The first replication experiment in 1992 determined the technique of producing tin metal in a home-made crucible (Yener and Vandiver 1993a, Yener 1994a) (Plate 14). Ground ore powder containing a low-grade 1% tin, taken from ground ore materials found in Early Bronze II/III pithouse floors, was used as the experimental charge. The first set of experimental crucibles were fabricated from local Celaller clays. Using a slab construction technique, three crucibles were made replicating some of the sizes and techniques of the actual archaeological crucibles. Enriched with one cup of water by vanning (panning with a shovel) (Plate 15), the ore was then placed in a homemade crucible made with local clay and chaff temper. The charge was placed in successive layers of charcoal and after twenty minutes of blowing through a single blowpipe, tin prills entrapped inside an envelop of glassy slag emerged inside the crucible (Plate 16a and b). During this experiment, tin metal prills (globules) encased in glassy slag were then released by grinding. The slag was thus in a powdery consistency and virtually invisible on survey unless microscale sampling methods were introduced.

In subsequent experiments (Yener and Earl 1994, Earl and Özbal 1996) the variables were altered considerably to determine the parameters of the process. Three separate qualities of charge were tested: a) a fine ground ore with relatively high tin content but unvannable because of iron contamination, b) ground ore as found in its original state without beneficiation with a vanning shovel, and c) a very small sample, enriched and placed into a microcrucible in a larger crucible imitating a bowl furnace and crucible. Other variables during these tests were the use of simultaneous blow pipes (up to three) (Plate 17), using the crucible with or without a cover, and the nature of the fuel used. The experiment with three blowpipes made the fire so hot that it melted the metal blowpipe, and vitrified the microcrucible. This indicated a temperature in excess of 1100° C. Variation in the charcoal affected the success of the smelt tremendously. The use of commercial charcoal briquettes resulted in an unsuccessful smelt in Cornwall, while wood charcoal completed the smelt efficiently and resulted in tin metal prills (globules). The test run utilizing a microcrucible was informative in providing information about crucible construction. Even though prills were produced, they penetrated the fabric of the microcrucible and were difficult to extract, unless the fabric was ground as well. This dramatically points out why the archaeological crucibles had a layer of dense, fine, well-levigated clay on the interior surface. The charge sample with high levels of iron that was not enriched fared poorly in vanning, thus making separation difficult.

The reconstruction of the Göltepe smelting stages, then, is based principally on the lack of furnaces and large-scale slag, the enormous quantities of vitrified crucible fragments of a distinct appearance, the composition of the ores, the tin-rich accretion on the crucibles, and the relative simplicity of the process. The second part of our model rests on the hypothesis that the vast amount of smelted tin was refined and melted in "melting" crucibles and then cast into bar-ingot-shaped molds for standardized ingots of tin metal. The bar ingots produced in these molds would have been suitable for measuring and transporting for alloying either at the site or at the urban centers. An alternative, semi-processed, ground, tin-rich smelted material could also have been transported.

Having produced small, sand-sized globules of tin metal and small amounts of slag, the next step was to attempt to make a tin bronze using this experimentally smelted material. This was accomplished at Cornwall using the experimental tin prills which had been manufactured in Turkey. While the tin in prill form could have been remelted and poured into a mold in order to produce an ingot, the alternative for alloying copper would be to add the prill-iron mixture to molten copper. The iron content of the tin produced in this manner would be rejected into a dross, producing a good bronze. This was successfully attempted in experimental conditions.

CHAPTER FIVE

CONCLUSIONS

The principal aim of this book has been to develop a new perspective on Anatolian metal studies by tracing the development of complex metal industries in the Taurus mountains. A much clearer picture of the history of the northern resource zones for Mesopotamia, Syria, and the Levant has emerged than was heretofore available. It is now evident, for example, that neither the development of prestate polities nor the emergence of complex urban centers in agriculturally fertile zones can be understood in isolation. Attempts to control access to needed highland metal resources and acquire advanced technology systems have been the rationale for a number of hypotheses involving the formation of Mesopotamia-induced colonial outposts in the latter part of the fourth millennium B.C., the Uruk period, and possibly earlier during the Ubaid period (late fifth-early fourth millennium). This theme of acquisition and control is further reinforced by the third millennium legends involving the military intervention in Anatolia of the Akkadian kings Sargon and Naram-Sin. The magnitude, if any, of these intrusions remains largely unknown since the archaeological history of the industrial sites located near the critical resources is only beginning to be determined. What appears likely is that Mesopotamian traders entered into an already complex environment of shifting and competing relationships between Anatolian city-states and vassals, highland metal producers, and agricultural enclaves. It may be postulated that some metal producers at times were embedded in a Syro-Mesopotamian exchange pattern. However, a multitude of alternatives, intra-Anatolian and with the Mediterranean, Black Sea/Caucasus, and Aegean regions, assured fluid economic relationships.

Information regarding one of the most important mining areas, the Taurus range, and metal workshops within reciprocating urban sites in Anatolia has helped fill in some of the gaps by articulating the impact of these incipient industrial processes on the local highland populations and exchange patterns in metal. Local production systems and the development of metallurgical technologies at sites in Anatolian resource centers were included in Chapter Two to determine whether they were in fact colonized, exploited, and receptive. Attention was called to the fact that enviable technological knowledge had germinated and accumulated in central and eastern Turkey, extending to the Caucasus, Balkans, and Iran. It is apparent that a set of technological and production styles had developed at

Chalcolithic period sites such as Mersin, Değirmentepe, Norşuntepe, the Amuq sites, Arslantepe, Tülintepe, and Tepecik prior to the arrival of Mesopotamian interests. Significantly, sophisticated copper-based metals and experimental alloys had appeared by the late 5th-4th millennium B.C. The production exploded in a vast array of alloys, stylistic types, decorations, and uses in pace with new outlets in Syro-Mesopotamia. Furthermore, metal absorbed the functions of prestige and power and became economically significant within localized traditions prior to and during the late 5th through 3rd millennia B.C. Innovations in the physical organization of the copper smelting industry had already gone beyond trinket manufacture.

Similar innovative changes have been tangible in the realm of material science and metallurgy. For example, complex two-piece molds for the casting and development of smelting crucibles attest to a growing production of critical importance to this region. Especially important in this regard have been the metal workshops at Norşuntepe and Değirmentepe, from which natural draft furnaces for smelting copper ores were recovered. Late Chalcolithic village sites in the eastern highlands such as Tülintepe, Tepecik, and Arslantepe have all yielded quantities of slag which suggests the smelting of sulfide and polymetallic ores. The earliest and most complete data sets for the study of alloying have been provided by the same sites. Değirmentepe and Mersin have yielded slag and artifacts, respectively, which document that arsenical bronzes were being made. Ternary bronzes, a combination of copper, arsenic, tin, or lead—perhaps as experimental alloys—, appear very early in Anatolia and continue to be used in the later periods. Occasionally examples of high zinc, antimony, or nickel levels have been found, perhaps a result of experimenting with polymetallic ores or impurities coming in through the use of flux. Arsenical copper (1% or higher As) was the first widely used alloy. Arsenic-rich copper objects of superior alloying dating to the Chalcolithic and Early Bronze Ages attest to the exploitation of richly colored, secondary sulfide ores. The impurities which can be measured in copper objects of the 5th millennium, indicate the widespread smelting of complex sulfide ores, surely an advance in metallurgy and specialization requiring skilled labor. The nascent emergence of tin bronze has been evident primarily in sites within the Amuq plain and Cilicia. In terms of administrative technology, Değirmentepe especially demonstrates a sophisticated distribution system using record-keeping devices such as seals. Both local and non-local styles in seals as well as sealings are represented at this site attesting to the storage and distribution of products manufactured in the households. On the other hand, at the later site of

Arslantepe, the locus of economic administrative functions was in larger-scale public buildings.

While site-centered production was the focus for the Chalcolithic period, by the early third millennium, special function industrial processing sites such as Göltepe and Kestel mine had appeared in the Taurus mountains. The wide range of metals, slags, vitrified products, and residues which were analyzed were part of the multistage metal production system. This data was utilized to recreate the technology of producing tin metal using replication experiments outlined in Chapter Four. The contention that tin production is untenable in Anatolia has proven incorrect. The ton of tin-rich vitrifed ceramics at Göltepe is assuredly testimony to that. The view that sophisticated metal technologies were brought into Anatolia (or other highland resource areas) is also not correct. The rich, detailed analytical evidence has clearly increased the confidence one may have in the explanations offered. Trace element analyses, microscopic research, metallurgical cross sections, lead isotope ratios, and microprobe, among other analytical techniques, have allowed us to define the physical properties of the metal industry. Thus the materials from the production sites of Göltepe and Kestel mine, discussed in Chapter Three, have been used inductively to give new insights into the unknown world of the technology and organization of specifically tin production. Placed within a wider cultural context, this technology is given new dimensions by attempts to localize technologies for other comparative purposes.

The development of metallurgy in Anatolia was an exceedingly complex process. The central Taurus region has shown that a multiplicity of metals were extracted from these sources from the earliest periods. Complex, organized, and metallurgically sophisticated industries became evident in the Chalcolithic period in the central and eastern Taurus mountain regions. Throughout most of their history, the lowlands and highlands were interconnected by traders and Bronze Age entrepreneurs. Recent investigations of these mining districts have revealed that a regional procurement strategy was already developed in the Early Bronze Age, one which tied together the mountain sources with the lowland markets. A two-tiered production system existed consisting of the sites which extracted ores, did the rough smelting, and cast the metal into ingots, and the urban centers which subsequently refined, crafted, and manufactured idiosyncratic metal items in workshops.

The work done in the central Taurus first-tier special-function sites has gone a long way towards couching intelligent questions regarding the context and organization of metal production in the region. By closing a significant gap in the understanding of metal production at a site within a strategic metal zone, research in the source zones has become central to forthcoming interpretative efforts seeking to pull together the growing corpus of metals from urban centers. In so doing, this investigation will illuminate the metallurgical development of a little-known region that was surely of fundamental importance to the entire ancient Near East.

BIBLIOGRAPHY

Adams, R. Mc., "Review of Larsen, M. T., The Old Assyrian City-State and Its Colonies," *Journal of Near Eastern Studies* 37 (1978) 265-69.

——, *Paths of Fire: An Anthropological Look at Technology* (Princeton 1996).

Adriaens, A., "Elemental Composition and Microstructure of Early Bronze Age and Medieval Tin Slags," *Mikrochimica Acta* 1124 (1996) 89-98.

Adriaens, A., K. A. Yener, and F. Adams, "Surface Analysis of Early Bronze Age Ceramic Crucibles from Göltepe, Turkey," *The Proceedings of the 6th European Conference on Applications of Surface and Interface Analysis*. H. J. Mathieu, B. Reihl and D. Briggs, eds. (Chichester and New York 1996) 123-26.

——, "An Analytical Study Using Electron and Ion Microscopy of Thin-Walled Crucibles from Göltepe, Turkey," *Journal of Archaeological Science* 26 (1999) 1069-1073.

Adriaens, A., K. A. Yener, F. Adams, and R. Levi-Setti, "SIMS Analyses of Ancient Ceramic Crucibles and Slags from Turkey," *Tenth International Conference on Secondary Ion Mass Spectrometry SIMS X*. A. Benninghoven, B. Hagenhoff and H. W. Werner, eds. (Chichester and New York 1997) 877-80.

Adriaens, A., B. Earl, H. Özbal, and K. A. Yener, "Tin Bronze Metallurgy in Transformation: Analytical Investigation of Crucible Fragments from Tell al-Judaidah, Amuq (Turkey) Dating to Circa 3000-2900 B.C.," *Proceedings of the 31st Archaeometry Symposium* (Budapest in press).

Adriaens, A., P. Veny, F. Adams, R. Sporken, P. Louette, B. Earl, H. Özbal, and K. A. Yener, "Analytical Investigation of Archaeological Powders from Göltepe (Turkey)," *Archaeometry* 41 (1999) 81-90.

Akay, E., and S. Uysal, "Post-Eocene Tectonics of the Central Taurus Mountains," *M.T.A. Bulletin* 108 (1988) 23-34.

Aksoy, B., "Prehistoric and Early Historic Pottery of the Bolkardağ Mining District," *XXXIV. Uluslararası Assiriyoloji Kongresi/Proceedings of the 34th International Assyriology Congress, July 6-10, 1987, Istanbul*. Hayat Erkanal *et al.*, eds. (Ankara 1998) 565-72.

Aksoy, B., and S. Duprés, "The Ceramic Assemblage from Göltepe," *Excavations at Göltepe*, Oriental Institute Publications. K. A. Yener, ed. (Chicago in prep).

Alessio, M., L. Allegri, C. Azzi, F. Bella, G. Calderoni, C. Cortesi, S. Improta, and V. Petrone, "^{14}C Dating of Arslantepe," *Origini* 12 (1983) 575-80.
Algaze, G., "The Uruk Expansion. Cross-cultural Exchange in Early Mesopotamian Civilization," *Current Anthropology* 30 (1989) 571-608.
——, *The Uruk World System: The Dynamics of Expansion of Early Mesopotamian Civilization* (Chicago 1993).
Algaze, G., A. Mısır, and T. J. Wilkinson, "Şanlıurfa Museum/University of California Excavations and Surveys at Titriş Höyük, 1991: A Preliminary Report," *Anatolica* 18 (1992) 33-60.
Algaze, G., R. Breuninger, and J. Knudstad, "The Tigris-Euphrates Archaeological Reconnaissance Project: Final Reports on the Birecik and Carchemish Dam Survey Areas," *Anatolica* 20 (1994) 1-96.
Alimov, K., N. Boroffka, and M. Bubnova, "Prähistorischer Zinnbergbau in Mittelasien," *Eurasia Antiqua* 4 (1998) 137-199.
Alkım, U. B., "The Amanus Region in Turkey. New Light on the Historical Geography and Archaeology," *Archaeology* 22 (1969) 280-89.
Alp, S., *Zylinder- und Stempelsiegel aus Karahöyük bei Konya* (Ankara 1968).
Andrews, P., "Excavating Mines (Copa Hill, Chinflon, Kestel)," *Bulletin of the Peak District Mines Historical Society* 12/3 (1994) 13-21.
Archi, Alfonso, "Bronze Alloys in Ebla," *Between the Rivers and Over the Mountains: Archaeologica Anatolica et Mesopotamica Alba Palmieri Dedicata.* M. Frangipane, H. Hauptmann, M. Liverani, P. Matthiae and M. Mellink, eds. (Rome Gruppo Editoriale Internazionale 1993) 615-25.
Arsebük, G., "Tülintepe: Some Aspects of a Prehistoric Village," *Beiträge zur Altertumskunde Kleinasiens.* R. M. Boehmer and H. Hauptman, eds. (Mainz 1983) 51-58.
——, "An Assemblage of Microlithic Engravers from the Chalcolithic Levels of Değirmentepe (Malatya)," *Jahrbuch für Kleinasiatische Forschung (Anadolu Arastırmaları)* X (in Memoriam Prof. Dr. U.B. Alkım) (1986) 132-42.
Ayhan, A., "Genetic Comparison of Lead-zone Deposits of Central Taurus," *Geology of the Taurus Belt. Proceedings 26-29 September 1983* (Ankara 1984) 335-42.
Bachmann, H.-G., *The Identification of Slags from Archeological Sites* (London 1982).
Bar-Adon, P., *The Cave of the Treasure* (Jerusalem 1980).

Bates, D., and S. Lees, "The Role of Exchange in Productive Specialization," *American Anthropologist* 79 (1977) 824-41.
Basalla, G., *The Evolution of Technology* (Cambridge 1988).
Begemann, F., E. Pernicka, and S. Schmitt-Strecker, "Metal Finds from Ilıpınar and the Advent of Arsenical Copper," *Anatolica* 20 (1994) 203-19.
Behm-Blancke, M., "Hassek Höyük. Vorläufiger Bericht über die Ausgrabungen der Jahre 1978-80," *Istanbuler Mitteilungen* 34 (1981) 5-93.
———, "Hassek Höyük. Vorläufiger Bericht über die Ausgrabungen der Jahre 1981-83," *Istanbuler Mitteilungen* 31 (1984) 31-149.
Belli, O., "Neue Funde Steinerner Gussformen aus Akçadağ bei Malatya," *Between the Rivers and Over the Mountains: Archaeologica Anatolica et Mesopotamica Alba Palmieri Dedicata*. M. Frangipane, H. Hauptmann, M. Liverani, P. Matthiae and M. Mellink, eds. (Roma 1993) 605-14.
Bilgi, Ö., "Metal Objects from Ikiztepe in Turkey," *Beiträge zur Allgemeinen und Vergleichenden Archäologie* 6 (1984) 31-96.
———, "Metal Objects from İkiztepe-Turkey," *Beiträge zur Allgemeinen und Vergleichenden Archäologie* 9/10 (1990) 119-219.
Binford, L. R., "General Introduction," *For Theory Building in Archaeology: Essays on Faunal Remains, Aquatic Resources, Spatial Analysis and Systematic Modeling*. L. R. Binford, ed. (New York 1977) 1-10.
———, "Researching Ambiguity: Frames of Reference and Site Structure," *Method and Theory for Activity Area Research*. S. Kent, ed. (New York 1987) 449-512.
Binford, L. R., and J. A. Sabloff, "Paradigms, Systematics, and Archaeology," *Journal of Anthropological Research* 38 (1982) 137-53.
Bishop, R. L., V. Canouts, S. P. De Atley, and P. L. Crown, "Sensitivity, Precision, and Accuracy: Their Roles in Ceramic Compositional Data Bases," *American Antiquity* 55 (1990) 250-70.
Bishop, R. L., and F. W. Lange, "Introduction," *The Ceramic Legacy of Anna O. Shepard*. R. L. Bishop and F. W. Lange, eds. (Niwot, CO 1991) 1-8.
Bishop, R. L., R. L. Rands, and G. R. Holly, "Ceramic Composition Analysis in Archaeological Perspective," *Archaeological Method and Theory*, Vol. 5. (New York 1982) 275-330.
Bittel, K., "Der Depotfund von Soloi-Pompeiopolis," *Zeitschrift für Assyriologie* 46 (1940) 183-205.
———, "Einige Kleinfunde aus Mysien und aus Kilikien," *Istanbuler Mitteilungen* 6 (1955) 113-18.

Blackman, M. J., "The Provenience of Obsidian Artifacts from Late Chalcolithic Levels at Aphrodisias," *Prehistoric Aphrodisias*, Vol. I. M. S. Joukowsky, ed. (Louvain, Belgium 1986) 279-87.
Blegen, C. W., *Troy I. General Introduction. The First and Second Settlements* (Princeton 1950).
Blegen, C. W., L. Caskey, and P. Rawson, *Troy II. The Third, Fourth and Fifth Settlements* (Princeton 1951).
Blumenthal, M. M., *Yüksek Bolkardağın Kuzey Kenar Bölgelerinin ve Batı Uzantılarının Jeolojisi/The Geology of the North and West Slope Areas of Bölkardağı* (Ankara 1956).
Bordaz, J., "The Suberde Excavations, Southwestern Turkey, An Interim Report," *Türk Arkeoloji Dergisi* 17/2 (1969) 43-71.
Bostanci, E., "The Mesolithic of Beldibi and Belbaşı and the Relation with other Findings in Anatolia," *Antropoloji* 3 (1965) 91-148.
Braidwood, R. J., *Mounds in the Plain of Antioch: An Archaeological Survey*, Oriental Institute Publications 48 (Chicago 1937).
Braidwood, R. J., and L. Braidwood, eds., *Excavations in the Plain of Antioch I: The Earlier Assemblages Phases A-J*, Oriental Institute Publications 61 (Chicago 1960).
——, *Prehistoric Village Archaeology in South-Eastern Turkey. The Eighth Millennium B.C. Site at Çayönü: Its Chipped and Ground Stone Industries and Faunal Remains* (Oxford 1982).
Braidwood, R. J., J. E. Burke, and N. H. Nachtrieb, "Ancient Syrian Coppers and Bronzes," *Journal of Chemical Education* 28 (1951) 87-96.
Braidwood, R. J., H. Çambel, C. H. Redman, and P. J. Watson, "Beginnings of Village-Farming Communities in Southeastern Turkey," *Proceedings of the National Academy of Sciences* 68 (1971) 1236-40.
R. J. Braidwood, L. S. Braidwood, B. Howe, C. A. Reed, and P. J. Wilson, *Prehistoric Archaeology along the Zargos Flanks*, Oriental Institute Publications 105 (Chicago 1983).
Brinkman, J. A., "Textual Evidence for Bronze in Babylonian in the Early Iron Age, 1000-509 B.C.," *Bronzeworking Centres of Western Asia c. 1000-539 B.C.* J. Curtis, ed. (London 1988) 135-68.
British Naval Intelligence Report, *A Handbook of Asia Minor Vol. IV, part 2. Cilicia, Anti-taurus and North Syria* (London 1919).
——, *Geographical Handbooks. Turkey*, Vol. 1 (London 1942).
Bromley, A., "Mineralogy of Tin Ores and Processing Products Kestel Mine Site and Göltepe Celaller, Anatolia, Turkey" (Unpublished report 1992).

Brooks, A., and J. Yellen, "The Preservation of Activity Areas in the Archaeological Record: Ethnoarchaeological and Archaeological Work in Northwest Ngamiland, Botswana," *Method and Theory for Activity Area Research*. S. Kent, ed. (New York 1987) 63-106.

Brumfiel, E., and T. K. Earle, *Specialization, Exchange and Complex Societies* (Cambridge 1987).

Çağatay, A., Y. Altun, and B. Arman, "Bolkardağ Sulucadere (Ulukışla-Niğde) Kalay içerikli Çinko-kurşun cevherleşmesinin minerolojisi [Mineralogy of the Tin Bearing Bolkardağ Sulucadere (Ulukışla-Niğde) Lead-zinc Mineralization]," *Turkiye Jeoloji Bulteni/Geological Bulletin of Turkey* 32 (1989) 15-20.

Çağatay, A., B. Arman, and Y. Altun, "Madenbelenitepe (Soğukpinar-Keles-Bursa) stannitinin incelenmesi," *Jeoloji Mühendisliği Dergisi/Geological Engineering* 13 (1982) 23-26.

Çağatay, A., and N. Pehlivan, "Celaller (Niğde-Çamardı) Kalay Cevherlesmeşinin Mineralojisi [Mineralogy of the Celaller Niğde-Çamardı Tin Mineralisation]," *Jeoloji Mühendisliği Dergisi/Geological Engineering* 32-33 (1988) 27-31.

Caldwell, J. R., *Investigations at Tal-i-Iblis* (Illinois 1967).

Caldwell, J. R., and S. M. Shahmirzadi, *Tal-i-Iblis, The Kerman Range and the Beginnings of Smelting* (Illinois 1966).

Çambel, H., and R. J. Braidwood, "An Early Farming Village in Turkey," *Scientific American* 222 (1970) 50-56.

Çambel, H., R. J. Braidwood, *et al.*, *The Joint Istanbul-Chicago Universities' Prehistoric Research in Southeastern Anatolia*, Vol. I (Istanbul 1980).

Canby, J. V., "Early Bronze 'Trinket' Moulds," *Iraq* 27 (1965) 42-58.

Caneva, I., "Scavi a Mersin-Yumuktepe," *Orient Express* (1996) 5-7.

———, "9000 years ago Yumuktepe," *The Anniversary of the Excavations at Yumuktepe (1993-1997)*. K. Köroğlu, ed. (Istanbul 1998) 9-12.

Caneva, C., and C. Giardino, "Extractive Techniques and Alloying in Prehistoric Central Anatolia: Experimental Methods in Archaeometallurgy," *Archaeometry 94. The Proceedings of the 29th International Symposium on Archaeometry*. S. Demirci, A. M. Özer and G. D. Summers, eds. (Ankara 1996) 451-59.

Caneva, C., and A. M. Palmieri, "Metalwork at Arslantepe in Late Chalcolithic and Early Bronze I: The Evidence from Metal Analysis," *Origini* 12 (1983) 637-54.

Caneva, C., M. Frangipane, and A. M. Palmieri, "I metalli di Arslantepe nel quadro dei piu antichi sviluppi della metallurgia vicino-orientale," *Quaderni de 'La ricerca scientifica'* 112 (1976-1979) (1985) 115-37.

Caneva, C., A. M. Palmieri, and K. Sertok, "Mineral Analysis in the Malatya Area. Some Hypotheses," *IV. Arkeometri Sonuçları Toplantısı* (Ankara 1989) 39-48.

——, "Copper Ores in the Malatya Region and Smelting Experiments," *V. Arkeometri Sonuçları Toplantısı* (Ankara 1990).

——, "Archaeometallurgical Research in the Malatya Area," *IX. Araştırma Sonuçları Toplantısı* (Ankara 1992) 227-34.

Caneva, C, M. K. Sertok, and A. M. Palmieri, "Malatya Çevresindeki Bakır Cevherleri ve Ergitme Deneyleri," *VI. Arkeometri Sonuçları Toplantısı* (Ankara 1991) 1-11.

Çevikbaş, A., and Ö. Öztunalı, "[Ore Deposits in the Ulukışla-Çamardı Basin]" *Jeoloji Mühendisliği* 39 (1991) 22-40.

Charles, J. A., "The Coming of Copper and Copper-Base Alloys and Iron: A Metallurgical Sequence," *The Coming of the Age of Iron.* T. A. Wertime and J. D. Muhly, eds. (New Haven 1980) 151-81.

——, "Determinative Mineralogy in the Early Development of Metals," *Journal of the Historical Metallurgy Society* 28 (1994) 66-68.

Chernykh, E. N., *Ancient Metallurgy in the USSR. The Early Metal Age* (Cambridge 1992).

Childe, V. G., "The Axes from Maikop and Caucasian Metallurgy," *Liverpool Annals of Archaeology and Anthropology* 23 (1936) 113-19.

——, "Archaeological Ages as Technological Stages," *Journal of the Royal Anthropological Institute of Great Britain and Ireland* 74 (1944) 7-24.

——, *Man Makes Himself* (New York 1951).

Childes, T. S., "Iron as Utility or Expression: Reforging Function in Africa," *Metals in Society: Theory Beyond Analysis*, MASCA Research Papers in Science and Archaeology 8/II. R. M. Ehrenreich, eds. (Philadelphia 1991) 57-69.

Cleuziou, S., and T. Berthoud, "Early Tin in the Near East," *Expedition* 25 (1982) 14-19.

Craddock, P., "Three Thousand Years of Copper Alloys: From the Bronze Age to the Industrial Revolution," *Application of Science in Examination of Works of Art.* P. A. England and L. van Zelst, eds. (Boston 1985) 59-67.

——, *Early Metal Mining and Production* (Washington D.C. 1995).

Crawford, H. E. W., "The Problem of Tin in Mesopotamian Bronzes," *World Archeology* 6 (1974) 242-47.

——, *Sumer and the Sumerians* (Cambridge 1991).

Cribb, R., *Nomads in Archaeology* (Cambridge 1991).

Çukur, A., and Ş. Kunç, "Analysis of Tepecik and Tülintepe Metal Artifacts," *Anatolian Studies* 39 (1989) 113-20.

———, "Acemhöyük Bakır Buluntu Analizleri," *V. Arkeometri Sonuçları Toplantısı* (Ankara 1990) 33-39.

Daniel, G., *The Origins and Growth of Archaeology* (New York 1967).

De Atley, S. P., and R. L. Bishop, "Toward an Integrated Interface for Archaeology and Archaeometry," *The Ceramic Legacy of Anna O. Shepard*. R. L. Bishop and F. W. Lange, eds. (Niwot, Colorado 1991) 358-82.

de Jesus, P. S., *The Development of Prehistoric Mining and Metallurgy in Anatolia* (Oxford 1980).

Diamant, S., and J. Rutter, "Horned Objects in Anatolia and the Near East and Possible Connexions with the Minoan 'Horns of Consecration'," *Anatolian Studies* 19 (1969) 147-78.

Earl, B., "Melting Tin in the West of England: A Study of an Old Art," *Journal of the Historical Metallurgy Society* 19 (1985) 153-61.

———, "Tin Preparation and Smelting," *The Industrial Revolution in Metals*. J. Day and R. F. Tylecote, eds. (London 1991) 47-83.

———, "Tin from the Bronze Age Smelting Viewpoint," *Journal of the Historical Metallurgy Society* 28 (1994) 117-20.

Earl, B., and H. Özbal, "Early Bronze Age Tin Processing at Kestel-Göltepe," *Archaeometry 94. Proceedings of the 29th International Symposium on Archaeometry*. S. Demirci, A. M. Özer, and G. D. Summers, eds. (Ankara 1996) 447-49.

Edens, C., "Dynamics of Trade in the Ancient Mesopotamian World System," *American Anthropologist* 94 (1992) 118-39.

Ehrenreich, R. M., "Archaeometallurgy and Archaeology: Widening the Scope," *Recent Trends in Archaeometallurgical Research*, MASCA Research Papers in Science and Archaeology 8/I. P. D. Glumac, ed. (Philadelphia 1991) 55-62.

Ekholm, K., and J. Friedman, "'Capital' Imperialism and Exploitation In Ancient World Systems," *Power and Propaganda*. M. T. Larsen, ed. (Copenhagen 1979) 41-58.

Epstein, S. M., "Physical and Cultural Constraint of Innovation in the Late Prehistoric Metallurgy of Cerro Huaringa, Peru," *Materials Issues in Art and Archaeology III*. P. B. Vandiver, J. Druzik and G. S. Wheeler, eds. (Pittsburgh 1992) 747-56.

Esin, U., *Kuantitatif Spektral Analiz Yardımıyla Anadolu'da Başlangıcından Asur Kolonileri Çağına Kadar Bakır ve Tunç Madenciliği I* (Istanbul 1969).

———, "Tepecik Excavations, 1970," *Keban Project 1970 Activities* (Ankara 1972) 149-58.

———, "Tepecik Excavations, 1972," *Keban Project 1972 Activities* (Ankara 1976a) 109-46.

——, "Tülintepe Excavations, 1972," *Keban Project 1972 Activities* (Ankara 1976b) 147-74.

——, "Die Anfänger der Metallverwendung und Bearbeitung in Anatolien (7500-2000 v. Chr.)," *Les débuts de la metallurgie. IXe Congres, Union Internationale des Sciences Préhistoriques et Protohistoriques.* H. Müller-Karpe, ed. (Nice 1976c) 199-242.

——, "Tepecik ve Tülintepe Kazılarına ait Arkeometrik Araştırmaların Arkeolojik Açıdan Değirlendirilmesi," *Arkeometri Ünitesi Bilimsel Toplantı Bildirileri*, Vol. 2. (Ankara 1981a) 157-82.

——, "Değirmentepe Kazısı 1979," *II. Kazı Sonuçları Toplantısı* (Ankara 1981b) 91-99.

——, "1980 yılı Değirmentepe (Malatya) Kazısı Sonuçları," *III. Kazı Sonuçları Toplantısı* (Ankara 1981c) 39-41.

——, "Tepecik Excavations, 1974," *Keban Project 1974-5 Activities* (Ankara 1982a) 95-127.

——, "Tülintepe Excavations, 1974," *Keban Project 1974-5 Activities* (Ankara 1982b) 127-33.

——, "Die Kulturellen Beziehungen zwischen Ostanatolien sowie Syrien anhand einiger Grabungs-und Oberflachenfunde aus dem oberen Euphrattal im 4. Jt. v. Chr," *Mesopotamien und seine Nachbarn.* H. Nissen and J. Renger, eds. (Berlin 1982c) 13-21.

——, "Zur Datierung der vorgeschichtlichen Schichten von Değirmentepe bei Malatya in der östlichen Türkei," *Beiträge zur Altertumskunde Kleinasiens.* R. M. Boehmer and H. Hauptmann, eds. (Mainz 1983a) 175-90.

——, "Değirmentepe (Malatya) Kazısı 1981 Yılı Sonuçları," *IV. Kazı Sonuçları Toplantısı* (Ankara 1983b) 11-29.

——, "Arkeometrik Açıdan Değirmentepe (Malatya) Kazıları," *Arkeometri Ünitesi Bilimsel Toplantı Bildirileri III* (Ankara 1983c) 141-62.

——, "Tepecik, Tülintepe (Altınova-Elazığ), Değirmentpe (Malatya) Kazıları," *I. Arkeometri Ünitesi Bilimsel Toplantı Bildirileri* (Ankara 1984) 68-112.

——, "Değirmentepe Kazısı 1983 Raporu," *VI. Kazı Sonuçları Toplantısı* (Ankara 1985a) 11-29.

——, "Some Small Finds from the Chalcolithic occupation at Degirmentepe (Malatya) in Eastern Turkey," *Studi di Palenologia in Onore de Salvatore M. Puglisi.* M. Liverani, A. Palmieri, and R. Peroni, eds. (Rome 1985b) 253-63.

——, "Dogu Anadolu'ya ait bazi Prehistorik Curuf ve Filiz Analizeri," *Jahrbuch für Kleinasiatische Forschung* Anadolu Araştırmaları (in Memoriam Prof. Dr. U.B. Alkim) X (1986) 143-60.

―, "Tepecik ve Tülintepe (Altinova-Elazig) ait Bazi Metal Curuf Analizleri," *II. Arkeometri Sonuçları Toplantı Bildirileri* (Ankara 1987) 69-79.

―, "An Early Trading Center in Eastern Anatolia," *Anatolia and the Ancient Near East. Studies in Honor of Tahsin Özgüç.* K. Emre et al., eds. (Ankara 1989) 135-41.

― "Değirmentepe (Malatya) Kalkolitik Obeyd Evresi Damga Mühür ve Mühür Baskıları." X. *Türk Tarih Kongresi* (Ankara 1990) 47-56.

―, "Aşıklı-Hüyük (Kızıkaya-Aksaray) Kurtarma Kazısı 1989," *Türk Arkeoloji Dergisi* 19 (1991) 1-34.

―, "Early Copper Metallurgy at the Pre-Pottery Site of Aşıklı," *Readings in Prehistory. Studies Presented to Halet Çambel* (Istanbul 1995) 61-79.

Esin U., and G. Arsebük, "Tülintepe Excavations, 1971," *Keban Project 1971 Activities* (Ankara 1974) 149-54.

―, "1982 Yılı Değirmentepe (Malatya) Kurtarma Kazısı," *V. Kazı Sonuçları Toplantısı* (Ankara 1983) 71-79.

Esin U., and S. Harmankaya, "1984 Değirmentepe (Malatya) Kurtarma Kazısı," *VII. Kazı Sonuçları Toplantısı* (Ankara 1986) 53-86.

―, "1985 Değirmentepe (Malatya-Imamlı Köyü) Kurtarma Kazısı," *VIII. Kazı Sonuçları Toplantısı* I (Ankara 1987) 95-137.

―, "Değirmentepe (Malatya) Kurtarma Kazısı," *IX. Kazı Sonuçları Toplantısı I* (Ankara 1988) 79-125.

Esin, U., O. Birgül, and L. Yaffe, "Değirmentepe Keramiklerinin Eser Element Analizi [Trace Element Analysis of Pottery from Değirmentepe]," *V. Arkeometri Unitesi Bilimsel Toplant Bildirileri* (Ankara 1985) 50-60.

Esin, U., E. Biçakçı, M. Özbaşaran, N. B. Atlı, D. Berker, I. Yağmur, and A. K. Atlı, "Salvage Excavations at the Pre-Pottery site of Aşıklı Hüyük in Central Anatolia," *Anatolica* 17 (1991) 123-74.

Forbes, R. J., *Studies in Ancient Technology*, Vol. VII (Geology and Mining) (Leiden 1963).

―, *Studies in Ancient Technology*, Vol. VIII (Metallurgy) (Leiden 1964a).

―, *Studies in Ancient Technology*, Vol. IX (Metals) (Leiden 1964b).

Frangipane, M., "Early Developments of Metallurgy in the Near East," *Studi di Paletnologia in onore di S.M. Puglisi*. M. Liverani, A. Palmieri and R. Peroni, eds. (Rome 1985) 215-28.

―, "The 1990 Excavations at Arslantepe, Malatya," *XIII/I. Kazı Sonuçları Toplantısı* (Ankara 1992) 177-95.

―, "Local Components in the Development of Centralized Societies in Syro-Anatolian Regions," *Between the Rivers and Over the Mountains:*

Archaeologica Anatolica et Mesopotamica Alba Palmieri Dedicata. M. Frangipane, H. Hauptmann, M. Liverani, P. Matthiae and M. Mellink, eds. (Roma 1993a) 133-61.

——, "New Groups of Clay-Sealings from the 4th Millennium Levels of Arslantepe-Malatya," *Aspects of Art and Iconography: Anatolia and its Neighbors.* Studies in Honor of Nimet Özgüç. M. J. Mellink, E. Porada and T. Özgüç, eds. (Ankara 1993b) 191-200.

——, "A Fourth Millennium Temple/Palace Complex at Arslantepe-Malatya. North-South Relations and the Formation of Early State Societies in the Northern Regions of Southern Mesopotamia," *Paléorient* 23 (1998) 45-73.

Frangipane, M., and A. Palmieri, "Perspectives on Protourbanization in Eastern Anatolia: Arslantepe (Malatya). An Interim Report on 1975-1983 Campaigns, a Protourban Centre of the Late Uruk Period," *Origini* 12 (1983a) 287-454.

——, "Cultural Developments at Arslantepe at the Beginning of Third Millennium," *Origini* 12 (1983b) 523-73.

——, "Urbanization in Perimesopotamian Areas: The Case of Eastern Anatolia," *Studies in the Neolithic and Urban Revolutions.* L. Manzanilla, ed. (Oxford 1987) 295-318.

——, "Aspects of Centralization in the Late Uruk Period in Mesopotamian Periphery," *Origini* 14 (1989) 539-60.

Frangipane, M., H. Hauptmann, M. Liverani, P. Matthiae, and M. Mellink, eds., *Between the Rivers and Over the Mountains: Archaeologica Anatolica et Mesopotamica Alba Palmieri Dedicata* (Roma 1983).

Frankfort, H., "Sumerians, Semites and the Origin of Copper-Working," *Antiquaries Journal* (1928) 217-35.

Franklin, A. D., J. S. Olin, and T. A. Wertime, eds., *The Search for Ancient Tin* (Washington D.C. 1978).

French, D. H., "Excavations at Can Hasan," *Anatolian Studies* 12 (1962) 27-40.

——, "Excavations at Can Hasan," *Anatolian Studies* 13 (1963) 29-42.

Gale, N. H., Z. A. Stos-Gale, and G. R. Gilmore, "Alloy Types and Copper Sources of Anatolian Copper Alloy Artifacts," *Anatolian Studies* 35 (1985) 143-73.

Garelli, P., *Les Assyriens en Cappadoce* (Istanbul 1963).

Garstang, J., *Prehistoric Mersin* (Oxford Clarendon Press 1953).

Geselowitz, M. N., "Iron Production in Prehistoric Europe," *Journal of Metals* (June, 1988) 52-53.

Giles, D. L., and E. P. Kuijpers, "Stratiform Copper Deposit, Northern Anatolia, Turkey. Evidence for Early Bronze I (2800 B.C.) Mining Activity," *Science* 186 (1974) 823-25.
Godoy, R., "Mining: Anthropological Perspectives," *Annual Review of Anthropology* 14 (1985) 199-217.
Goldman, H., *Excavations at Gözlü Kule, Tarsus*, Vols. I-II (Princeton University Press 1956).
Griffitts, W. R., J. P. Albers, and O. Öner, "Massive Sulfide Copper Deposits of the Ergani Maden Area Southeastern Turkey," *Economic Geology* 6 (1972) 701-16.
Güterbock, H., "Ein neues Bruchstuck der Sargon-Erzählung "König des Schlacht"," *Mitteilungen der Deutsches Orient Gesellschaft* 101 (1969) 14-26.
Hall, M., and S. Steadman, "Anatolia and Tin: Another Look," *Journal of Mediterranean Archaeology* 4/1 (1991) 77-90.
Hallo, W. W., "From Bronze Age to Iron Age in Western Asia," *The Crisis Years: The 12th Century B.C.* W. A. Ward and M. S. Joukowsky, eds. (Dubuque Kendall/Hunt 1992) 1-9.
Hamilton, W. J., *Researches in Asia Minor, Pontus and Armenia I* (London 1842).
Hard, R. J., and K. A. Yener, "The Function of Anatolian Ground and Battered Tools" (Paper delivered SAA meetings, New Orleans 1991).
Hartmann, A., "Ergebnisse der spektralanalytischen Untersuchung aneolithischer Goldfunde aus Bulgarien," *Studia Präehistorica* (1978) 1-2.
Hartmann, A., and E. Sangmeister, "The Study of Prehistoric Metallurgy," *Angewandte Chemie* 11 (1972) 620-29.
Hauptmann, A., "The Earliest Periods of Copper Metallurgy in Feinan, Jordan," *Old World Archaeometallurgy*. A. Hauptmann, E. Pernicka, and G. A. Wagner, eds. (Bochum: Deutschen Bergbau-Museums 1989) 119-35.
——, "Developments in Copper Metallurgy During the 4th/3rd Millennium B.C. at Feinan/Jordan" (Abstracts of the British Museum Conference on the Prehistory of Mining and Metallurgy 13-18 September 1995).
Hauptmann, A., G. Weisgerber, and H.-G. Bachmann, "Ancient Copper Production in the Area of Feinan, Khirbet En-Nahas, and Wadi el-Jariye, Wadi Arabah, Jordan," *History of Technology: The Role of Metals*. S. J. Fleming and H. R. Schenck, eds. (1989) 7-16.
Hauptmann, A., J. Lutz, E. Pernicka, and Ü. Yalçın, "Zur Technologie der Frühesten Kupferverhüttung im Östlichen Mittelmeerraum," *Between the Rivers and Over the Mountains: Archaeologica Anatolica et*

Mesopotamica Alba Palmieri Dedicata. M. Frangipane, H. Hauptmann, M. Liverani, P. Matthiae and M. Mellink, eds. (Roma Gruppo Editoriale Internazionale 1993) 541-72.

Hauptmann, A., F. Begemann, E. Heitkemper, E. Pernicka, and S. Scmitt-Strecker, "Early Copper Produced at Feinan Wadi Araba, Jordan: The Composition of Ores and Copper," *Archeomaterials* 6 (1992) 1-33.

Hauptmann, H., "Die Grabungen auf dem Norşun-Tepe, 1970," *Keban Project 1970 Activities*, I, 3 (Ankara: Middle Eastern Technical University 1972) 103-22.

——, "Die Grabungen auf dem Norşun-Tepe, 1971," *Keban Project 1971 Activities*, I, 4 (Ankara: Middle Eastern Technical University 1974) 87-106.

——, "Die Entwicklung der Frühbronze-zeitlichen siedlung auf dem Norşuntepe in Ostanatolien," *Archäeologiches Korrespondenzblatt* 6 (1976a) 9-20.

——, "Die Grabungen auf dem Norşuntepe, 1972," *Keban Project 1972 Activities*, I, 5 (Ankara: Middle Eastern Technical University 1976b) 71-90.

——, "Die Grabungen auf dem Norşun-Tepe, 1974," *Keban Project 1971 Activities*, I, 7 (Ankara: Middle Eastern Technical University 1982) 41-70.

——, "Ein Kultgebäude in Nevali Çori," *Between the Rivers and Over the Mountains: Archaeologica Anatolica et Mesopotamica Alba Palmieri Dedicata.* M. Frangipane, H. Hauptmann, M. Liverani, P. Matthiae, and M. Mellink, eds. (Roma Gruppo Editoriale Interna 1993) 37-70.

Helms, M. W., *Craft and the Kingly Ideal. Art, Trade, and Power* (Austin: University of Texas Press 1993).

Heskel, D. L., "Early Bronze Age Anatolian Metal Objects: A Comparison of Two Techniques for Utilizing Spectographic Analyses," *Annali di Istituto Orientale di Napoli* 40 (1980) 473-502.

——, "A Model for the Adoption of Metallurgy in the Ancient Middle East," *Current Anthropology* 24 (1983) 362-66.

Heskel, D., and C. C. Lamberg-Karlovsky, "An Alternative Sequence for the Development of Metallurgy: Tepe Yahya, Iran," *The Coming of the Age of Iron.* T. A. Wertime and J. D. Muhly, eds. (New Haven: Yale University Press 1980) 229-65.

Hodder, I., "The Meaning of Discard. Ash and Domestic Space in Baringo," *Method and Theory for Activity Area Research.* S. Kent, ed. (New York: Columbia University Press 1987) 424-48.

——, "Excavations at Çatalhöyük," *Anatolian Archaeology* 1 (1995) 3-5.

Hong, S., J.-P. Candelone, C. C. Patterson, and C. F. Boutron, "History of Ancient Copper Smelting Pollution during Roman and Medieval Times Recorded in Greenland Ice," *Science* 272 (1996) 247-48.

Hosler, D., *The Sounds and Colors of Power. The Sacred Metallurgical Technology of Ancient West Mexico* (Cambridge, Ma.: M.I.T. Press 1994).

Hosler, D., H. Lechtman, and O. Holm, *Axe-Monies and Their Relatives* (Washington: Dumbarton Oaks 1990).

Johnson, G. A., "The Changing Organization of Uruk Administration on the Susiana Plain," *Archaeology of Western Iran*. F. Hole, ed. (Washington D.C.: Smithsonian Institution Press 1987) 107-39.

Jovanovic, B., "The Oldest Copper Metallurgy in the Balkans," *Expedition* (1978) 9-17.

———, "Primary Copper Mining and the Production of Copper," *Scientific Studies in Early Mining and Extractive Metallurgy*, British Museum Occasional Papers 20. P. T. Craddock, ed. (London 1980) 31-40.

Jovanovic, B., and B. S. Ottoway, "Copper Mining and Metallurgy in the Vinca Group," *Antiquity* 50 (1976) 104-11.

Junghans, J., E. Sangmeister, and M. Schröder, *Metallanalyse kupferzeitlicher und fruhbronzezeitlicher Bodenfunde aus Europe*, SAM vol. I (Berlin Gebr. Mann Verlag 1960).

———, *Kupfer und Bronze in der frühen Metallzeit Europas 1-3*, SAM vol. II (Berlin: Gebr. Mann Verlag 1968).

———, *Kupfer und Bronze in der frühen Metallzeit Europas 4* (Berlin: Gebr. Mann Verlag 1974).

Kaptan, E., "Ancient Miners' Shovels and Ore Carrier Discovered in Espiye-Bulancik Area," *M.T.A. Bulletin* 91 (1978) 99-110.

———, "The Significance of Tin in Turkish Mining History and its Origin," *M.T.A. Bulletin* 95/96 (1983) 164-72.

———, "New Discoveries in the Mining History of Turkey in the Neighborhood of Gümüşköy, Kutahya," *M.T.A. Bulletin* 97/98 (1984) 140-47.

———, "Ancient Mining in the Tokat Province Anatolia: New Finds," *Anatolica* 13 (1986) 19-36.

———, "Türkiye Madencilik Tarihine ait Çamardı-Celaller Köyü Yöresindeki Buluntular," *IV. Arkeometri Sonuçları Toplantısı* (Ankara: Department of Antiquities and Museums 1989) 1-16.

———, "Finds Relating to the History of Metallurgy in Turkey," *M.T.A. Bulletin* 111 (1990a) 75-84.

———, "Turkiye Madencilik Tarihine ait Celaller (Niğde) Yoresindeki Sarituzla-Göltepe Buluntulari," *V. Arkeometri Sonuçları Toplantısı* (Ankara: Department of Antiquities and Museums 1990b) 13-32.

——, "Tin and Ancient Tin Mining in Turkey," *Anatolica* 21 (1995) 197-203.
Kaptan, E. and M. Yildirim, "A Double Function Mineral Dressing Device," *29th International Symposium on Archaeometry, May 1-14, 1994, Ankara* (Ankara 1995) 571-74.
Kent, S., "Activity Areas and Architecture: An Interdisciplinary View of the Relationship Between the Use of Space and Domestic Built Environments," *Domestic Architecture and the Use of Space*. S. Kent, ed. (Cambridge University Press 1990).
Killick, D., "The Relevance of Recent African Iron-Smelting Practice to Reconstruction of Prehistoric Smelting Technology," *Recent Trends in Archaeometallurgical Research*, MASCA Research Papers in Science and Archaeology 8/I. P. D. Glumac, ed. (Philadelphia: University of Pennsylvania 1991) 47-54.
Kohl, P. L., "The Use and Abuse of World Systems Theory: The Case of the Pristine West Asian State," *Advances in Archaeological Method and Theory*, Vol. 11. M. B. Schiffer, ed. (San Diego: Academic Press 1987) 1-36.
Koşay, H. Z., *Ausgrabungen von Alaca Hüyük, 1936*, Türk Tarih Kurumu V. no. 2a (Ankara 1944).
——, *Alaca Hüyük Kazısı* (*Les Fouilles d'Alaca Hüyük Rapport préliminaire sur les travaux en 1937-39*), Türk Tarih Kurumu Series V. no. 5 (Ankara 1951).
——, *Keban Project Pulur Excavations 1968-1970*, Middle Eastern Technical University series II, no. 1 (Ankara 1976).
Koşay, H. Z., and M. Akok, *Ausgrabungen von Alaca Hüyük: Vorbericht uber die Forschungen und Entdeckungen von 1940-48*, Türk Tarih Kurumu Series V. no. 6 (Ankara 1966).
——, *Alaca Hüyük Excavations 1963-1967*, Türk Tarih Kurumu V. no. 28 (Ankara 1973).
Koşay, H. Z., and K. Turfan, "Erzurum-Karaz Kazısı Raporu," *Belleten* 23 (1959) 349-413.
Kramer, C., *Village Ethnoarchaeology: Rural Iran in Archaeological Perspective* (New York: Academic Press 1982).
Kristiansen, K., "From Stone to Bronze: The Evolution of Social Complexity in Northern Europe, 2300-1200 B.C.," *Specialization, Exchange and Complex Societies*. E. Brumfiel and T. K. Earle, eds. (Cambridge University Press 1987) 30-51.
Kühne, H., *Die Keramik von Tell Chuera und ihre Beziehungen zu Funden aus Syrien-Palätina, der Türkei und dem Iraq* (Berlin: Gebr. Mann Verlag 1976).

Kunç, Ş., and A. Çukur, "Bakır Buluntularda iz Element Dağılımı," *III Arkeometri Sonuçları Toplantısı* (Ankara: Ministry of Culture 1988a) 97-105.

——, "Tepecik ve Tülintepe Buluntularnn Eser Element Analizleri," *III Arkeometri Sonuçları Toplantısı* (Ankara: Ministry of Culture 1988b) 77-96.

Kunç, Ş., A. Eker, S. Kapur, and N. Gündoğdu, " Değirmentepe Curuf Buluntulari Analizi," *IV Arkeometri Unitesi Bilimsel Toplantı Bildirileri* (Ankara: Tübitak 1984) 26-30.

——, "Değirmentepe Curuf Buluntu Analizeri II," *VI Arkeometri Unitesi Bilimsel Toplantı Bildirileri* (Ankara: Tübitak 1986) 114-20.

——, "Degirmentepe Curuf Analizleri III," *II Arkeometri Sonuçları Toplantısı* (Ankara: Ministry of Culture 1987).

Ladame, G., *Bolkardağ Madeni*, M.T.A. unpublished report number 304 (Ankara 1938).

Lamb, W., "Excavations at Kusura near Afyon Karahisar," *Archaeologia* 86 (1936) 1-64.

Larsen, M. T., *The Old Assyrian City-State and its Colonies*, Copenhagen Studies in Assyriology Vol. 4 (Copenhagen: Akademisk Forlag 1976).

——, "Commercial Networks in the Ancient Near East," *Centre and Periphery in the Ancient World*. M. Rowlands, M. T. Larsen, and K. Kristiansen, eds. (Cambridge: Cambridge University Press 1987) 47-56.

Laughlin, G. J., Microscopy and Microanalysis of Tin Ore Processing Evidence from the Early Bronze Age. PhD in Metallurgy and Materials Engineering, Illinois Institute of Technology (1998).

Lechtman, H., "A Metallurgical Site Survey in the Peruvian Andes," *Journal of Field Archaeology* 3 (1976) 1-42.

——, "Traditions and Styles in Central Andean Metalworking," *The Beginning of the Use of Metals and Alloys. Papers from the Second International Conference on the Beginning of the Use of Metals and Alloys, Zhengzhou, China, 11-26 October 1986* R. Maddin, ed. (Cambridge, Ma.: M.I.T. 1988) 344-78.

——, "The Production of Copper-arsenic Alloys in the Central Andes: Highland Ores and Coastal Smelters?" *Journal of Field Archaeology* 18 (1991) 43-76.

Lechtman, H., and A. Steinberg, "The History of Technology: An Anthropological Point of View," *The History and Philosophy of Technology*. G. Bugliarello and D. Doner, eds. (Chicago: University of Illinois Press 1979) 135-60.

Lemonnier, P., "Bark Capes, Arrowheads and Concorde: On Social Representations of Technology," *The Meaning of Things: Material*

Culture and Symbolic Expression. I. Hodder, ed. (London: Unwin Hyman 1989) 156-71.

Lemonnier, P. *Technological Choices. Transformation in Material* (New York: Routledge 1993).

Limet, H., *Le Travail du métal au pays de Sumer au temps de la IIIe dynastie d'Ur*, Bibliothèque de la Fac. de philos. et lettres de l'Université de Liège Fascicle CLV (Paris 1960).

——, "Les métaux a l'époque d'Agade," *Journal of the Economic and Social History of the Orient* 15 (1972) 3-24.

Lutz, J., E. Pernicka, and G. A. Wagner, "Chalkolithische Kupferverhüttung in Murgul, Ostanatolien," *Handwerk und Technologie im Alten Orient. Ein Beitrag zur Geschichte der Technik im Altertum*. R.-B. Wartke, ed. (Mainz Verlag Philipp von Zabern 1994) 59-66.

Lyon, J. D., " Intra-Site and Interregional Organization of Metal Production at Değirmentepe," unpublished paper, Metal Technology and Society Organizational Seminar, University of Chicago 1997.

Maden Tetkik Arama Enstitusu (M.T.A.), *Iron Ore Deposits of Turkey*, M.T.A. no. 118 (Ankara 1964).

——, *Arsenic, Antimony and Gold Deposits of Turkey*, M.T.A. no. 129 (Ankara 1970).

——, *Lead, Copper, and Zinc Deposits of Turkey*, M.T.A. no. 133 (Ankara 1972).

——, *Geology of the Taurus Belt* (Ankara: M.T.A. 1984).

Maddin, R., T. Stech, and J. D. Muhly, " Distinguishing Artifacts Made of Native Copper," *Journal of Archaeological Science* 7 (1980) 211-25.

——, "Çayönü Tepesi. The Earliest Archaeological Metal Artifacts," *Découverte du Métal*. J.-P. Mohen and C. Éluère, eds. (Paris: Picard 1991) 375-86.

Majidzadeh, Y., "An Early Prehistoric Coppersmith Workshop at Tepe Ghabristan," *Archaeologische Mitteilungen aus Iran* 6 (1976) 82-93.

Mallowan, M. E. L., "Excavations at Brak and Chagar Bazar," *Iraq* 9 (1947) 1-259.

Marfoe, L., "Cedar Forest to Silver Mountain: Social Change and the Development of Long-Distance Trade in Early Near Eastern Societies," *Centre and Periphery in the Ancient World*. M. J. Rowlands, M. T. Larsen and K. Kristiansen, eds. (Cambridge University Press 1987) 25-35.

Maxwell-Hyslop, R., *Western Asiatic Jewellery, c. 3000-612 B.C.* (London: Methuen 1971).

Mellaart, J., "Preliminary Report on a Survey of Pre-Classical Remains in Southern Turkey," *Anatolian Studies* 4 (1954) 175-240.
——, "Archaeological Survey of the Konya Plain," *Anatolian Studies* 9 (1959) 31-33.
——, "Early Cultures of the South Anatolian Plateau," *Anatolian Studies* 11 (1961) 159-84.
——, "Excavations at Çatal Hüyük, 1961. First Preliminary Report," *Anatolian Studies* 12 (1962) 41-65.
——, "Excavations at Çatal Hüyük, 1962. Second Preliminary Report," *Anatolian Studies* 13 (1963a) 43-103.
——, "Early Cultures of the Anatolian Plateau, II. The Late Chalcolithic and Early Bronze Age in the Konya Plain," *Anatolian Studies* 11 (1963b) 199-236.
——, "Excavations at Çatal Hüyük, 1963. Third Preliminary Report," *Anatolian Studies* 14 (1964) 39-119.
——, *Earliest Civilizations of the Near East* (London: Thames and Hudson 1965).
——, "Excavations at Çatal Hüyük, 1965, Fourth Preliminary Report," *Anatolian Studies* 16 (1966) 165-91.
——, *Çatal Hüyük. A Neolithic Town in Anatolia* (New York McGraw-Hill 1967).
——, *Excavations at Hacılar I, II* (Edinburgh British Institute of Archaeology at Ankara Publications 1970).
Mellink, M.J., "Anatolian and Foreign Relations of Tarsus," *Anatolia and the Ancient Near East. Studies in Honor of Tahsin Özgüç.* K. Emre, et al. eds. (Ankara: Türk Tarih Kurumu Yayinlari 1989) 319-331.
Mokyr, J., *The Level of Riches. Technological Creativity and Economic Progress* (Oxford University Press 1990).
Moorey, P. R. S., "The Archaeological Evidence for Metallurgy and Related Technologies in Mesopotamia c. 5500-2100 B.C.," *Iraq* 44 (1982) 13-38.
——, *Materials and Manufacture in Ancient Mesopotamia: The Evidence of Archaeology and Art, Metals and Metalwork, Glazed Materials and Glass*, B.A.R. International Series 237 (Oxford 1985).
——, *Ancient Mesopotamian Materials and Industries. The Archaeological Evidence* (Oxford: Clarendon Press 1994).
Muhly, J. D., *Copper and Tin. The Distribution of Mineral Resources and the Nature of the Metal Trade in the Bronze Age*, Transactions of the Connecticut Academy of Arts Vol. 43 (Hamden CT: Archon Books 1973).
——, *Supplement to Copper and Tin*, Transactions Vol. 46 (1976).

——, "New Evidence for Sources of and Trade in Bronze Age Tin," *The Search for Ancient Tin.* A. D. Franklin et al., eds. (Washington D.C.: Smithsonian Institution Press 1978) 43-48.

——, "The Beginnings of Metallurgy in the Old World," *The Beginning of the Use of Metals and Alloys. Papers from the Second International Conference on the Beginning of the Use of Metals and Alloys, Zhengzhou, China, 11-26 October 1986.* R. Madden, ed. (Cambridge, Ma: M.I.T. Press 1988) 2-20.

——, "Çayönü Tepesi and the Beginnings of Metallurgy in the Old World," *Old World Archaeometallurgy.* A. Hauptmann, E. Pernicka, and G. A. Wagner, eds. (Bochum: Deutschen Bergbau-Museums 1989) 1-13.

——, "Early Bronze Age Tin and the Taurus," *American Journal of Archaeology* 97/2 (1993) 239-53.

Muhly, J. D., F. Begemann, Ö. Öztunali, E. Pernicka, S. Schmitt-Strecker, and G. A. Wagner, " The Bronze Age Metallurgy of Anatolia and the Question of Local Tin Sources," *Archaeometry '90.* E. Pernicka and G. Wagner, eds. (Switzerland: Birkhauser Verlag 1991) 209-20.

Müller-Karpe, A., *Altanatolisches Metallhandwerk.* Offa-Bücher Band 75. (Neumünster: Wachholtz Verlag 1994).

Müller-Karpe, H., "Zur Definition und Benennung chronologischer Stufen der Kupferzeit, Bronzezeit und älteren Eisenzeit," *Jahresbericht des Instituts für Vorgeschichte der Universität Frankfurt a.M.* (1974) 7-18.

Müller-Karpe, M., *Metallgefäße im Iraq I. Von den Anfängen bis zur Akkad-Zeit*, Prähistorische Bronzefunde Abteilung II, Band 14 (Stuttgart: Franz Steiner Verlag 1993).

Nash, J., *We Eat the Mines* (New York: Columbia University Press 1979).

Natalja, R., I. Ginda, and K. Vera, "Copper Production from Sulfide Polymetal Ores of the East-Northern Balkans Eneolithic Culture," *Program and Abstracts, International Symposium on Archaeometry May 20-24, 1996* (University of Illinois at Urbana-Champaign 1996).

Neuninger, H., R. Pittioni, and W. Siegl, "Frühkeramikzeitliche Kupfergewinnung in Anatolien," *Archaeologia Austriaca* 35 (1964) 98-110.

Nissen, H.G., *The Early History of the Ancient Near East 9000-2000 B.C.* (University of Chicago Press 1988).

Northover, J. P., "Properties and Use of Arsenic-Copper Alloys," *Old World Archaeometallurgy.* A. Hauptmann, E. Pernicka, and G. A. Wagner, eds. (Bochum: Deutschen Bergbau-Museums 1989) 111-18.

Oates, J., "Trade and Power in the Fifth and Fourth Millennia B.C.: New Evidence from Northern Mesopotamia," *World Archaeology* 24/3 (1993) 403-22.

Orlin, L., *Assyrian Colonies in Cappadocia* (The Hague: Mouton 1970).

Özbal, H., "Tepecik ve Tülintepe Metal, Filiz ve Curuf Analizleri Sonuçları," *Arkeometri Unitesi Bilimsel Toplantısı Bildirileri III* (Ankara: Tubitak 1983) 203-18.

———, "Değirmentepe Metal, Curuf ve Filiz Analizleri," *Arkeometri Unitesi Bilimsel Toplantısı Bildirileri VI* (Ankara: Tubitak 1986) 101-13.

———, "Kestel-Göltepe Kalay Işletmeleri," *VIII. Arkeometri Sonuçları Toplantısı* (Ankara: Directorate General of Monuments and Museums 1993) 303-13.

———, "Early Metal Technology at Hacınebi Tepe," *Anatolica* 23 (1997) 139-143.

Özbal, H., and H. Ibar, "Galena ve Sulucadere Stannit Filizi Uzerinde Izabe ve Kupelasyon Deneyleri [Smelting and Cupellation Experiments on Galena and Sulucadere Stannite Ore]," *Aksay Ünitesi Bilimsel Toplantı Bildirileri I* (Ankara: Tubitak 1990) 187-94.

Özbal, H., B. Earl, and A. Mieke Adriaens, "Early Fourth Millennium Copper Metallurgy at Hacınebi," *Anatolica* 24 (1998) 167-70.

Özbal, H., B. Earl, A. Adriaens, and B. Gedik, "Metal Sources of Ikiztepe: Surveys at Merzifon-Durağan Region" (Paper delivered at the 21st International Symposium of Excavations, Surveys and Archaeometry May 24-28, 1999, Ankara Turkey).

Özdoğan, A., "Life at Çayönü During the Pre-Pottery Neolithic Period (according to the artifactual assemblage)," *Readings in Prehistory. Studies Presented to Halet Çambel* (Istanbul: Prehistorya Yayinlari/Graphis Yayinlari 1995) 79-100.

Özdoğan, M., *Lower Euphrates Basin 1977 Survey*, Middle Eastern Technical University I, no. 2 (Ankara 1977).

———, "Archaeological Evidence on Early Metallurgy and Lime Processing at Çayönü Tepesi. (Paper delivered at the Early Metallurgy Symposium 1995, Bochum, Germany, in press).

Özgüç. T., "Report on a Work-shop belonging to the Late Phase of the Colony Period Ib," *Belleten* 73 (1955) 77-80.

———, "Fraktin Kabartması Yanındaki Prehistorik Ev," *Anatolia* 1 (1956) 58-64.

———, "New Observations on the Relationship of Kültepe with Southeast Anatolia and North Syria during the Third Millennium B.C.," *Ancient Anatolia: Aspects of Change and Cultural Development. Essays in Honor of Machteld J. Mellink.* J. V. Canby, E. Porada, B. Ridgway, and T. Stech, eds. (Madison: University of Wisconsin Press 1986) 31-47.

Özguç, T., and M. Akok, "Objects from Horoztepe," *Belleten* 21 (1957) 211-19.

Özten, A., "A Group of Early Bronze Age Pottery from the Konya and Niğde Regions," *Anatolia and the Ancient Near East. Studies in Honor of Tahsin Özgüç*. K. Emre et al., eds. (Ankara: Turk Tarih Kurumu Yayınları 1989) 407-18.

Palmieri, A., "Scavi nell área sud-occidentale di Arslantepe," *Origini* 7 (1973a) 55-228.

——, "Arslantepe (Malatya). Report on the Excavations 1971-1972," *Türk Arkeoloji Dergisi* 21 (1973b) 137-46.

——, "Scavi ad Arslantepe (Malatya)," *Quaderni de 'La ricerca scientifica'* 100 (1978) 311-75.

——, "Excavations at Arslantepe (Malatya)," *Anatolian Studies* 31 (1981) 101-19.

——, "Excavations at Arslantepe, 1983," *VI. Kazı Sonuçları Toplantısı* (Ankara: Ministry of Culture 1984) 71-78.

——, "Eastern Anatolia and Early Mesopotamian Urbanization: Remarks on Changing Relations," *Studi di paletnologie in onore di Salvatore M. Puglisi*. M. Liverani, A. Palmieri, and R. Peroni, eds. (Roma: Università degli Studi di Roma "La Sapienza" 1985) 191-213.

——, "1984 Excavations at Arslantepe," *VII. Kazı Sonuçları Toplantısı* (Ankara: Ministry of Culture 1986) 29-36.

——, "The 1985 Campaign at Arslantepe, Malatya," *VIII. Kazı Sonuçları Toplantısı* (Ankara: Ministry of Culture 1987) 67-74.

——, "Storage and Distribution at Arslantepe-Malatya in the Late Uruk Period," *Anatolia and the Ancient Near East. Studies in Honor of Tahsin Özgüç*. K. Emre, B. Hrouda, M. Mellink, and N. Özgüç, eds. (Ankara: Türk Tarih Kurumu Basımevi 1989) 419-30.

Palmieri, A., and K. Sertok, "Minerals in and Around Arslantepe," *IX. Arkeometri Sonuçları Toplantısı* (Ankara: Ministry of Culture 1994) 119-52.

Palmieri, A., A. Hauptmann, and K. Hess, "The Metal Objects in the 'Royal' Tomb Dating from 3000 B.C. Found at Arslantepe (Malatya): a New Alloy (Cu-Ag)," *XIII. Arkeometri Sonuçları Toplantısı* (Ankara: Ministry of Culture 1998) 115-21.

Palmieri, A., K. Sertok, and E. Chernykh, "Archaeometallurgical Research at Arslantepe," *VIII. Arkeometri Sonuçları Toplantısı* (Ankara: Ministry of Culture 1993a) 391-98.

——, "From Arslantepe Metalwork to Arsenical Copper Technology in Eastern Anatolia," *Between the Rivers and Over the Mountains:*

Archaeologica Anatolica et Mesopotamica Alba Palmieri Dedicata. M. Frangipane, H. Hauptmann, M. Liverani, P. Matthiae and M. Mellink, eds. (Roma: Gruppo Editoriale Internazionale 1993b) 573-99.
Palmieri, A., A. Hauptmann, K. Hess, and K. Sertok," The Composition of Ores and Slags found at Arslantepe, Malatya," *Archaeometry 94. The Proceedings of the 29th International Symposium on Archaeometry.* S. Demirci, A. M. Özer and G. D. Summers, eds. (Ankara: Tübitak 1996) 447-49.
Palmieri, A. M., A. Hauptmann, K. Hess, and K. Sertok, "1995 Archaeometallurgical Campaign at Arslantepe," *XII. Arkeometri Sonuçları Toplantısı* (Ankara: Turkish Ministry of Culture 1997) 57-63.
Palmieri, A. M., A. Hauptmann, K. Sertok, and K. Hess, "Archaeometallurgical Surveys in 1994 at Malatya-Arslantepe and Its Surroundings," *XI. Arkeometri Sonuçları Toplantısı* (Ankara: Turkish Ministry of Culture 1995) 105-15.
Patterson, C. C., "Native Copper, Silver, and Gold Accessible to Early Metallurgists," *American Antiquity* 36 (1971) 288-321.
Pehlivan, N. A., and T. Alpan, *Niğde Masifi Altın-kalay Cevherleşmesi ve Ağır Mineral Çalışmaları ön Raporu* "Preliminary Reprot of the Gold/tin Mineralization and Heavy Mineral Research at the Niğde Massif," M.T.A. unpublished report (Ankara 1986).
Penhallurick, R. D., *Tin in Antiquity* (London 1987).
Pernicka, E., F. Begemann, C. Schmitt-Strecker, and P. P. Grimanis, "On the Composition and Provenance of Metal Artefacts from Poliochni on Lemnos," *Oxford Journal of Archaeology* 9 (1990) 263-97.
Pfaffenberger, B., "Fetishised Objects and Humanised Nature: Towards an Anthropology Technology," *Man* (N.S.) 23 (1988) 236-52.
——, "Social Anthropology of Technology," *Annual Review of Anthropology* 21 (1992) 491-516.
Pigott, V. C., "Discussion of Metals in Society," *Metals in Society: Theory Beyond Analysis*, MASCA Research Papers in Science and Archaeology 8/ II. R. M. Ehrenreich, ed. (Philadelphia 1991) 81-84.
——, "Near Eastern Archaeometallurgy: Modern Research and Future Directions," *The Study of the Ancient Near East in the 21st Century: The William Foxwell Albright Centennial Conference.* J. S. Cooper and G. M. Schwartz, eds. (Winona Lake, IN 1996) 139-76.
Pigott, V. C., W. Rostoker, and J. Dvorak, "From the Crucible to Copper in Prehistoric Thailand: A Preliminary Reconstruction of the Production Process" (Paper read at the International Symposium on Archaeometry, Toronto, 16-20 May 1988).

Pigott, V. C., and S. Natapintu, "Archaeological Investigations into Prehistoric Copper Production: The Thailand Archaeometallurgy Project 1984-1986," *The Beginnings of the Use of Metals and Alloys*. R. Maddin, ed. (Cambridge, Ma: MIT Press 1988) 156-62.

Powell, M., "Identification and Interpretation of Long Term Price Fluctuations in Babylon: More on the History of Money in Mesopotamia," *Altorientalische Forschungen* 17 (1990) 76-99.

Prag, K., "Silver in the Levant in the Fourth Millennium B.C.," *Archaeology in the Levant. Essays for Kathleen Kenyon*. R. Moorey and P. Parr, eds. (Warminster: Aris and Phillips 1978) 36-45.

Rapp, G., Jr., "Native Copper and the Beginning of Smelting: Chemical Studies," *Early Metallurgy in Cyprus, 4000-500 B.C.* J. D. Muhly, R. Maddin and V. Karageorghis, eds. (Nicosia: Cyprus Department of Antiquities 1982) 32-40.

Rapp, G., Jr., J. Zhychun, and R. Rothe, "Using Instrumental Neutron Activation Analysis (INAA) to Source Ancient Tin (Cassiterite)," *Abstracts of the International Symposium on Archaeometry, May 20-24 1996* (Urbana: University of Illinois 1996) 87.

Renfrew, C., "Obsidian in Western Asia: A Review," *Problems in Economic and Social Archaeology*. G. Sieveking et al., eds. (London: Duckworth 1977) 137-50.

——, "Varna and the Emergence of Wealth in Prehistoric Europe," *The Social Life of Things, Commodities in Cultural Perspective*. A. Appadurai, ed. (Cambridge University Press 1986) 141-68.

Rice, P. M., "Evolution of Specialized Pottery Production: A Trial Model," *Current Anthropology* 22 (1981) 219-40.

——, "Economic Change in the Lowland Maya Late Classic Period," *Specialization, Exchange and Complex Societies*. E. Brumfiel and T. K. Earle, eds. (Cambridge University Press 1987) 76-85.

——, "Specialization, Standardization, and Diversity: A Retrospective," *The Ceramic Legacy of Anna O. Shepard*. R. L. Bishop and F. W. Lange, eds. (Niwot University of Colorado 1991) 257-79.

Rosenberg, M., "Hallan Çemi Tepesi: Some Further Observations Concerning Stratigraphy and Material Culture," *Anatolica* 20 (1994) 121-40.

Rostoker, W., and J. R. Dvorak, "Some Experiments with Co-Smelting of Copper Alloys," *Archeomaterials* 5 (1991) 5-20.

Rostoker, W., V. C. Pigott, and J. R. Dvorak, "Direct Reduction to Copper Metal by Oxide-Sulfide Mineral Interaction," *Archeomaterials* 3 (1989) 69-87.

Rothenberg, B., *Timna: Valley of the Biblical Copper Mines* (London: Thames & Hudson 1972).

―, *The Egyptian Mining Temple at Timna* (London: Institute of Archaeology 1988).

―, *The Ancient Metallurgy of Copper* (London: University College 1990).

Ryan, C. W., *A Guide to the Known Minerals of Turkey* (Ankara: M.T.A. 1960).

Sagona, A.G. *The Caucasian Region in the Early Bronze Age*, Vols I-III, BAR International Series 214. (Oxford 1984).

Sasson, J. M., "Instances of Mobility among Mari Artisans," *Bulletin of the American Schools of Oriental Research* 190 (1968) 46-55.

Sayre, E. V., K. A. Yener, E. C. Joel, and I. L. Barnes, "Statistical Evaluation of the Presently Accumulated Lead Isotope Data from Anatolia and Surrounding Regions," *Archaeometry* 34 (1992) 73-105.

Seton-Williams, M. V., "Cilician Survey," *Anatolian Studies* 4 (1954) 121-74.

Schliemann, H., *Ilios. City and Country of the Trojans* (London: John Murray 1881).

Schmandt-Besserat, D., "Ocher in Prehistory: 300,000 Years of the Use of Iron Ores as Pigments," *The Coming of the Age of Iron*. T. A. Wertime and J. D. Muhly, eds. (New Haven: Yale University Press 1980) 127-49.

―, *Before Writing: From Counting to Cuneiform* (Austin, Texas 1992).

Schmidt, Klaus, "Frühneolithische Tempel. Ein Forschungsbericht zum präkeramischen Neolithikum Obermesopotamiens," *Mitteilungen des Deutsches Orient Gesellschaft* 130 (1998) 17-49

Schmitt-Strecker, C., F. Begemann, and E. Pernicka, "Chemische Zusammensetzung und Bleiisotopenverhältnisse der Metallfunde von Hassek Höyük," *Hassek Höyük Naturwissenschaftliche Untersuchungen und lithische Industrie*. M.R. Behm-Blancke, ed. (Tübingen: Ernst Wasmuth Verlag 1992) 108-23.

Schneider, J., "Was There a Pre-Capitalist World-System?" *Peasant Studies* 6 (1977) 20-29.

Selimkhanov, I. R., "Ancient Tin Objects of the Caucasus and the Results of their Analysis," *The Search for Ancient Tin*. A. D. Franklin et al., eds. (Washington D.C.: Smithsonian Institution Press 1978) 53-58.

Serdaroğlu, U., *Surveys in the Lower Euphrates Basin*, Middle Eastern Technical University no. 1 (Ankara 1977).

Seton-Williams, M. V., "Cilician Survey," *Anatolian Studies* 4 (1954) 121-74.

Seymour, D., and M. Schiffer, "A Preliminary Analysis of Pithouse Assemblages from Snaketown, Arizona," *Method and Theory for*

Activity Area Research. S. Kent, ed. (New York: Columbia University Press 1987) 549-603.
Shalev, S., and P. J. Northover, "The Metallurgy of the Nahal Mishmar Hoard Reconsidered," *Archaeometry* 35 (1993) 35-47.
Sharp, W. E., and S. K. Mittwede, "Was Kestel Really the Source of Tin for Ancient Bronze," *Geoarchaeology* 9 (1994) 155-58.
Sharpless, E., "Mercury Mines at Konia," *Engineering and Mining Journal* Sept 26 (1908) 601-3.
Shimada, I., "Pre-History Metallurgy and Mining in the Andes: Recent Advances and Future Tasks," *Quest of Mineral Wealth: Aboriginal and Colonial Mining and Metallurgy in Spanish America.* A. K. Craig and R. C. West, eds. *Geoscience and Man* 33 (Baton Rouge, LA 1990) 37-73.
Shimada, I., and J. F. Merkel, "Copper-Alloy Metallurgy in Ancient Peru," *Scientific American* 265 (1991) 80-86.
Silistreli, U., "Koşk Höyük Figurin ve Heykelcikleri," *Belleten* 207-8 (1990) 497-504.
Smith, C. S., "Analysis of the Copper Bead from Ali Kosh," *Prehistory and Human Ecology of the Deh Luran Plain*, Memoirs of the Museum of Anthropology no. 1. F. Hole, ed. (Ann Arbor: University of Michigan 1969) 427ff.
——, *A Search for Structure: Selected Essays on Science, Art and History* (Boston: Massachusetts Institute of Technology Press 1981).
Solecki, R., "A Copper Mineral Pendant from Northern Iraq," *Antiquity* 43 (1969) 311-14.
Sperl, G., "Urgeschichte des Bleis," *Zeischrift für Metallkunde* 81/11 (1990) 799-801.
Steadman, S., and M. Hall, "Anatolia and Tin: Another Look," *Journal of Mediterranean Archaeology* 4/1 (1991) 217-34.
Stech, T., "Neolithic Copper Metallurgy in Southwest Asia," *Archeomaterials* 4 (1990) 55-61.
Stech, T. and V. C. Pigott, "The Metals Trade in Southwest Asia in the Third Millennium B.C.," *Iraq* 48 (1986) 39-64
Stein, G. J., "On the Uruk Expansion," *Current Anthropology* 31 (1990) 66-67.
——, "Economy, Ritual and Power in Ubaid Mesopotamia," *Chiefdoms and Early States in the Near East.* G. J. Stein and M. Rothman, eds. (Madison, WI: Prehistory Press 1994) 35-46.
Stein, G. J., et al., "Uruk Colonies and Anatolian Communities: An Interim Report on the 1992-1993 Excavations at Hacinebi, Turkey," *American Journal of Archaeology* 100 (1996) 205-60.

Steponaitas, V., "Settlement Hierarchies and Political Complexity in Nonmarket Societies: The Formative Period in the Valley of Mexico," *American Anthropologist* 83 (1981) 320-63.

Stòs-Gale, Z., N. H. Gale, and G. R. Gilmore, "Early Bronze Age Trojan Metal Sources and Anatolians in the Cyclades," *Oxford Journal of Archaeology* 3/3 (1984) 23-43.

Stronach, D. B., "The Development and Diffusion of Metal Types in Early Bronze Anatolia," *Anatolian Studies* 7 (1957) 90-125.

Summers, G., "Chalcolithic Pottery from Kabakulak (Niğde) collected by Ian Todd," *Anatolian Studies* 41 (1991) 125-32.

Tallon, F., with J.-M. Malfoy and M. Menu, *Metallurgie susienne. I. De la fondation de Suse au XVIIIe siecle avant J.-C.* (Paris: Editions de la Runion des Muses nationaux 1987).

Taylor, J. W., "Erzgebirge Tin: A Closer Look," *Oxford Journal of Archaeology* 2 (1983) 295-97.

———, "Yugoslavian Tin Deposits and the Early Bronze Age Industries of the Aegean Region," *Oxford Journal of Archaeology* 6 (1987) 287-300.

Thomsen, C. J., *Ledetraad til Nordisk Oldkyndighed/A Guide to Northern Antiquities* (Copenhagen 1836). Translated by Lord Ellesmere (London 1848) 63-68.

Time-Life Editors, *Anatolia: Cauldron of Cultures* (Alexandria, VA: Time-Life Books 1995).

Tobler, A. J., *Excavations at Tepe Gawra*, Vol. I and II (Philadelphia: University Museum 1950).

Todd, I. A., "Aşıklı Hüyük. A Protoneolithic Site in Central Anatolia," *Anatolian Studies* 16 (1966) 139-63.

———, "The Dating of Aşıklı Hüyük in Central Anatolia," *American Journal of Archaeology* 72 (1968) 157-58.

Trigger, B. G., "Archaeology at the Crossroads: What's New?" *Annual Review of Anthropology* 13 (1984) 275-300.

Turner, F. J., *The Frontier in American History* (New York: Holt and Co. 1920).

Tylecote, R. F., *Early Metallurgy in the Near East* (London: The Metals Society 1970).

———, "Can Copper be Smelted in a Crucible?" *Journal of the Historical Metallurgy Society* 8 (1974) 54.

———, *A History of Metallurgy* (London: The Metals Society 1976).

———, "Chalcolithic Metallurgy in the Eastern Mediterranean," *Chalcolithic Cyprus and Western Asia*, British Museum Occasional Papers no. 26. Julian Reade, ed. (London 1981) 41-52.

——, *The Early History of Metallurgy in Europe* (London: Longman 1987).

Tylecote, R. F., and P. J. Boydell, "Experiments on Copper Smelting Based on Early Furnaces Found at Timna," *Archaeometallurgy* 1 (1978) 27-51.

Tylecote, R. F., H. A. Ghaznavi, and P. J. Boydell, "Partitioning of Trace Elements Between the Ores, Fluxes, Slags and Metal During the Smelting of Copper," *Journal of Archaeological Science* 4 (1977) 305-33.

Tylecote, R. F., E. Photos, and B. Earl, "The Composition of Tin Slags from the South-west of England," *World Archaeology* 20 (1989) 434-50.

Vandiver, P. B., K. A. Yener, and L. May, "Third Millennium B.C. Tin Processing Debris from Göltepe (Anatolia)," *Materials Issues in Art and Archaeology III*. P. B. Vandiver, J. Druzik and G. S. Wheeler, eds. (Pittsburgh: Materials Research Society 1992) 545-69.

Vandiver, P. B., R. Kaylor, J. Feathers, M. Gottfried, K. A. Yener, W. F. Hornyak, and A. Franklin, "Thermoluminescence Dating of a Crucible Fragment from an Early Tin Processing Site in Turkey," *Archaeometry* 35 (1993) 295-98.

van Driel, G., "Seals and Sealings from Jebel Aruda 1974-1978," *Akkadica* 33 (1983) 34-62.

van Loon, M., *Korucutepe. Final Report on the Excavations of the Universities of Chicago, California (Los Angeles) and Amsterdam in the Keban Reservoir, Eastern Anatolia*, Vol. 2 (Amsterdam 1978).

Waetzoldt, H., "Zur Terminologie der Metalle in den Texten aus Ebla," *La Lingua di Ebla*, Instituto U. Orientale no. XIV (Napoli 1981) 364-78.

Waetzoldt, H., and H. Hauptmann, *Wirtschaft un Gesellschaft von Ebla* (Heidelberg: Heidelberger Orientverlag 1989).

Wagner, G. A., Ö. Öztunalı, and C. Eibner," Early Copper in Anatolia," *Old World Archaeometallurgy*. A. Hauptmann, E. Pernicka and G. A. Wagner, eds. (Bochum: Deutschen Bergbau-Museums 1989) 299-305.

Wagner, G. A., E. Pernicka, T. C. Seeliger, I. B. Lorenz, F. Begemann, S. Schmitt-Strecker, C. Eibner, and Ö. Öztunalı, "Geochemische und isotopische Charakteristika früher Rohstoffquellen für Kupfer, Blei, Silber und Gold in der Türkei," *Jahrbuch des Römisch-Germanischen Zentralmuseums Mainz* 33 (1986) 723-52.

Waldbaum, J. C., *From Bronze to Iron. The Transition from the Bronze Age to the Iron Age in the Eastern Mediterranean*, Studies in Mediterranean Archaeology 54 (Göteborg 1978).

——, "Copper, Iron, Tin, Wood: The Start of the Iron Age in the Eastern Mediterranean," *Archeomaterials* 3 (1989) 111-22.

Wallerstein, I., *The Modern World System*, Vol. 1 (New York: Academic Press 1974).
——, *The Modern World System*, Vol. 2 (New York: Academic Press 1980).
——, *The Modern World System*, Vol. 3 (New York: Academic Press 1989).
Waterbolk, H. T., and J. J. Butler, "Comments on the Use of Metallurgical Analysis in Prehistoric Studies, I. A Graph Method for the Grouping and Comparison of Quantitative Spectro-Analyses of Prehistoric Bronzes," *Helinium* 5 (1965) 227-51.
Watkins, T., "Cultural Parallels in the Metalwork of Sumer and North Mesopotamia in the Third Millennium B.C.," *Iraq* 45 (1983) 18-23.
Weiss, H., *The Origins of Cities in Dry-Farming Syria and Mesopotamia in the Third Millennium B.C.* (Connecticut: Four Quarters Publishing 1986).
Weiss, H. et al., "The Genesis and Collapse of 3rd Millennium North Mesopotamian Civilization," *Science* 261 (1993) 995-1004
Wells, P. S., *Farms, Villages, and Cities: Commerce and Urban Origins in Late Prehistoric Europe* (Ithaca: Cornell University Press 1984).
Wertime, T. A., "Man's First Encounters with Metallurgy," *Science* 146 (1964) 1257-67.
——, "The Beginnings of Metallurgy: A New Look," *Science* 182 (1973) 875-87.
——, "The Search for Ancient Tin: The Geographic and Historic Boundaries," *The Search for Ancient Tin*. A. D. Franklin et al., eds. (Washington D.C.: Smithsonian Institution 1978) 1-6.
——, "Pyrotechnology: Man's Fire Using Crafts," *Early Technologies*. D. Schmandt-Besserat, ed. (Malibu: Undena 1979).
Wertime, T. A. and J. D. Muhly, eds., *The Coming of the Age of Iron* (New Haven: Yale University Press 1980).
Whallon, R., *An Archaeological Survey of the Keban Reservoir Area of East-Central Turkey*, Memoirs of the Museum of Anthropology University of Michigan no. 11 (Ann Arbor, MI 1979).
Wilcke, C., *Das Lugalbanda epos* (Wiesbaden: Harrassowitz 1969).
Wilkinson, T. J., "The Structure and Dynamics of Dry-Farming States in Upper Mesopotamia," *Current Anthropology* 35 (1994) 483-520.
——,"The History of the Lake of Antioch: A Preliminary Note," *Festschrift for Michael Astour*. R. E. Averbeck, ed. (Bethesda, MD: 1998) 557-76.
Willies, L., "An Early Bronze Age Tin Mine in Anatolia," *Bulletin of the Peak District Mines Historical Society* 11 (1990) 91-96.

——, "Report on the 1991 Archaeological Survey of Kestel Tin Mine, Turkey," *Bulletin of the Peak District Mines Historical Society* 11 (1991) 241-47.

——, "Reply to Hall and Steadman," *Journal of Mediterranean Archaeology* 5/1 (1992) 99-103.

——, "Reply to J.D: Muhly, 'Early Bronze Age Tin and the Taurus'," *American Journal of Archaeology* 97 (1993) 262-64.

——, "Firesetting Technology," *Mining Before Powder, Bulletin of the Peak District Mines Historical Society* 12/3. T. D. Ford and L. Willies, eds. (1994) 1-8.

——, "Kestel Tin Mine, Turkey. Interim Report 1995," *Bulletin of the Peak District Mines Historical Society* 12/5 (1995) 1-11.

Woolley, C. L., "Hittite Burial Customs," *Liverpool Annals of Archaeology and Anthropology* 6 (1914) 87-98.

——, *The Royal Cemetery. Ur Excavations II*, Vol. I and II (London and Philadelphia: Trustees of the Two Museums 1934).

——, *Ur Excavations IV: The Early Periods* (London and Philadelphia 1956).

Woolley, C. L., et al., *Carchemish. Report on the Excavations at Jerablus on Behalf of the British Museum, Part III: The Excavations in the Inner Town* (Oxford University Press 1952).

Wright, H. T., "The Evolution of Civilizations," *American Archaeology Past and Future*. D. J. Meltzer et al., eds. (Washington D.C.: Smithsonian Institution Press 1986) 323-68.

Yakar, J., "Regional and Local Schools of Metalwork in Early Bronze Age Anatolia, Part I," *Anatolian Studies* 34 (1984) 59-86.

——, "Regional and Local Schools of Metalwork in Early Bronze Age Anatolia, Part II," *Anatolian Studies* 35 (1985) 25-38.

Yalçın, Ü., "Der Keulenkopf von Can Hasan (TR): Naturwissenschaftliche Untersuchung und neue Interpretation," *Metallurgica Antiqua* 8. T. Rehren, A. Hauptmann and J. D. Muhly, eds. (Bochum: Der Anschnitt 1998) 279-89.

Yener, K. A., "Third Millennium B.C. Interregional Trade in Southwest Asia with Special Reference to the Keban Region of Turkey" (Ph.D. diss., Columbia University 1980).

——, "The Archaeometry of Silver in Anatolia: The Bolkardağ Mining District," *American Journal of Archaeology* 90 (1986) 469-72.

——, "Niğde-Çamardı'nda Kalay Buluntuları," *IV. Arkeometri Sonuçları Toplantısı* (Ankara: General Directorate of Antiquities 1989) 17-28.

——, "Arkeometri Projesi: Çamardı 1988 Çalışmaları," *V. Arkeometri Sonuçları Toplantısı* (Ankara: General Directorate of Antiquities 1990) 1-12.

——, "1990 Göltepe, Niğde Kazısı," *XIII. Kazı Sonuçları Toplantısı* (Ankara: General Directorate of Antiquities and Museums 1992) 275-89.

——, "Göltepe Kazısı 1991 Sezonu," *XIV. Kazı Sonuçları Toplantısı* (Ankara: General Directorate of Antiquities and Museums 1993) 231-47.

——, "Managing Metals: An Early Bronze Age Tin Production Site at Göltepe, Turkey," *The Oriental Institute News and Notes* 140 (1994a) 1-4.

——, "Göltepe/Kestel 1992," *XV. Kazı Sonuçları Toplantısı* (Ankara: General Directorate of Antiquities and Museums 1994b) 201-9.

——, Review of R. Maddin, ed., *The Beginning of the Use of Metals and Alloys* (Massachusettes 1988) *Journal of Near Eastern Studies* 53 (1994c) 301-4.

——, "Göltepe 1993 Kazı Sonuçları," *XVI. Kazı Sonuçları Toplantısı* (Ankara: General Directorate of Antiquities and Museums 1995a) 177-88.

——, "Swords, Armor, and Figurines: A Metalliferous View from the Central Taurus," *Biblical Archaeologist* 58 (Dedicated to Peter Neve) (1995b) 41-47.

——, "Göltepe and Kestel," *Civilizations of the Ancient Near East*, Vol III. J. Sasson, ed. (New York: Scribners 1995c) 1519-21.

——, "The Archaeological Season at Göltepe, Turkey, 1994," *The Oriental Institute 1994-1995 Annual Report* (1995d) 22-28.

——, "Göltepe, Kestel, Taurus Mountains, Turkey," *The Encyclopedia of Near Eastern Archaeology*. E. M. Meyers, ed. (Oxford: University Press 1996a) 155-56, 283-84, 425-26.

——, "Excavations at Kestel Mine, Turkey," *The Oriental Institute 1996-1997 Annual Report* (Chicago: The Oriental Institute 1997a) 58-61.

——, "An Early Bronze Age Mortuary Chamber Inside the Kestel Tin Mine: The 1996 Excavation Season," *The Oriental Institute News and Notes* 152 (1997b) 5-7.

Yener, K. A., and B. Earl, "Replication Experiments of Tin Using Crucibles, Göltepe 1992," *IX. Arkeometri Sonuçları Toplantısı* (Ankara: General Directorate of Antiquities 1994) 163-76.

Yener, K. A., and M. Goodway, "Response to Mark E. Hall and Sharon R. Steadman, 'Tin and Anatolia: Another Look'," *Journal of Mediterranean Archaeology* 5 (1992) 77-90.

Yener, K. A., and H. Özbal, "Tin in the Turkish Taurus Mountains: The Bolkardağ Mining District," *Antiquity* 61 (1987) 64-71.

Yener, K. A., and A. Toydemir, "Byzantine Silver Mines: An Archaeometallurgy Project in Turkey," *Ecclesiastical Silver Plate in*

Sixth-Century Byzantium. S. A. Boyd and M. M. Mango, eds., (Washington D.C., Dumbarton Oaks: 1993) 155-68.

Yener, K. A., and P. B. Vandiver, "Tin Processing at Göltepe, an Early Bronze Age Site in Anatolia," *American Journal of Archaeology* 97 (1993a) 207-37.

Yener, K. A., P. Jett, and M. Adrieans, "Silver and Copper Artifacts from Ancient Anatolia," *Journal of Metals* 45/5 (1995) 70-72.

Yener, K. A., and P. B. Vandiver, with Appendix by L. Willies, "Reply to J.D. Muhly, 'Early Bronze Age Tin and the Taurus'," *American Journal of Archaeology* 97 (1993b) 255-64.

Yener, K. A., H. Özbal, A. Minzoni-Deroche, and B. Aksoy, "Bolkardağ: Archaeometallurgy Surveys in the Taurus Mountains, Turkey," *National Geographic Research* 5/3 (1989a) 477-94

Yener, K. A., H. Özbal, E. Kaptan, A. N. Pehlivan, and M. Goodway, "Kestel: An Early Bronze Age Source of Tin Ore in the Taurus Mountains, Turkey," *Science* 244 (1989b) 200-3.

Yener, K. A., H. Özbal, B. Earl, and A. Adriaens, "The Analyses of Metalliferous Residues, Crucible Fragments, Experimental Smelts, and Ores from Kestel Tin Mine and the Tin Processing Site of Göltepe," *Proceedings of the Conference of Ancient Mining and Metallurgy,* British Museum Occasional Publications. P. Craddock, ed. (London: in press).

Yener, K. A., E. V. Sayre, E. Joel, H. Özbal, I. L. Barnes, and R. H. Brill, "Stable Lead Isotope Studies of Central Taurus Ore Sources and Related Artifacts from Eastern Mediterranean Chalcolithic and Bronze Age Sites," *Journal of Archaeological Science* 18 (1991) 541-77.

Yener, K. A., T. Wilkinson, S. Branting, E. Friedman, J. Lyon, and C. Reichel, "The 1995 Oriental Institute Amuq Regional Projects," *Anatolica* 22 (1996) 49-84.

Young, T. C. Jr., P. E. L. Smith, and P. Mortensen, eds., *The Hilly Flanks and Beyond: essays on the prehistory of southwestern Asia presented to Robert J. Braidwood, November 15, 1982* (Chicago 1983).

Zaccagnini, C., "Patterns of Mobility among Ancient Near Eastern Craftsmen," *Journal of Near Eastern Studies* 42 (1983) 245-64.

——, "The Transition from Bronze to Iron in the Near East and the Levant: Marginal Notes," *Journal of the American Oriental Society* 110 (1990) 493-502.

Zwicker, U., "Investigations on the Extractive Metallurgy of Cu/Sb/As Ore and Excavated Smelting Products from Norşun-Tepe (Keban) on the Upper Euphrates (3500-2800 B.C.)," *Aspects of Early Metallurgy,*

British Museum Occasional Papers 17. W. A. Oddy, ed. (London 1977) 13-26.

——, "Untersuchungen zur Herstellung von Kupfer und Kupferlegierungen im Bereich des östlichen Mittelmeeres (3500-1000 v. Chr.)," *Old World Archaeometallurgy*. A. Hauptmann, E. Pernicka and G. A. Wagner, eds. (Bochum Deutschen Bergbau-Museums 1989) 191-201.

——, "Natural Copper-arsenic Alloys and Smelted Arsenic Bronzes in Early Metal Production," *Découverte du Métal*. J.-P. Mohen and C. Éluère, eds. (Paris: Picard 1991) 331-40.

Zwicker, U., H. Greiner, K.-H. Hofmann, and M. Reitinger, "Smelting, Refining and Alloying of Copper and Copper Alloys in Crucible Furnaces during Prehistoric up to Roman Times," *Furnaces and Smelting Technology in Antiquity*, British Museum Occasional Paper no 48. P. T. Craddock and M. J. Hughes, eds. (London 1985) 103-15.

INDEX

Acemhöyük 102
adz 65
adz, bronze 46
adz, copper 66
Aegean 5, 10, 68, 125
Afghanistan 71, 75
Africa 9
agate 68
Ahlatlibel 67
Ai Bunar 89
Alaca Höyük 67, 68, 69
Aladağ 76
Ali Kosh 6
Alişar 62, 67, 68, 69
Alkım, H. 46
Alkım, U. 46
alloy 4, 6, 23 n. 1, 28, 29, 41, 45, 55, 66, 107, 126
alloying 4, 8, 18, 29, 32, 33, 41, 52, 53, 56, 63, 67, 69, 109, 123, 126
alloying material 39, 58, 68, 72, 73, 75, 121
Altınova valley 26, 32, 57, 61, 62
aluminosilicates 118
aluminum 118, 119
Amanus 2
Amuq 26, 30, 31, 32, 45, 51, 58, 60, 62 n. 9, 65, 74, 84, 102, 126
Amuq phase A 25
Amuq phase B 25
Amuq phase D 26, 34, 61, 63, 66
Amuq phase E 26, 34, 44, 62, 63, 65
Amuq phase F 34, 44, 58, 62, 64
Amuq phase G 47, 50 n. 5, 51
Amuq phase G-related pottery 60
Amuq phase H 50 n. 5, 51 n. 8, 75
Amuq phase I 75
Amuq phase J 51, 67
Anatolia 1, 2, 3, 4, 5, 6, 10, 11, 12, 14, 15, 17, 19, 22, 23, 25, 27, 28, 30, 31, 32, 33, 44, 45, 48, 49, 51, 65, 67, 68, 69, 70, 71, 72, 74, 75, 76, 80, 81, 82, 84, 85, 92, 102, 105, 108, 125, 126, 127
Anatolian clinky metallic ware 87, 95, 102, 106
anglesite 77
annaku 75
annealing 3, 21, 22, 23, 29, 52, 69
annealing twins 21, 22, 23
Antalya 18
antimony 28, 55, 63, 126
antimony-rich ore 55

Antitaurus 2, 33
apatite 80
Arabian plateau 30
archaeological surveys 76, 78-79, 81-85
archaeometallurgical surveys 28, 41
Arsebük, G. 33
arsenic 1, 21, 23, 28, 29, 32, 33, 40, 45, 46, 52, 53, 54, 55, 56, 59, 62, 63, 64, 65, 66, 67, 69, 117, 118, 126
arsenic-antimony-lead-nickel alloy 57
arsenic mineral 29, 52, 53, 55, 57, 59, 62
arsenic-lead-antimony ore 56
arsenic-nickel rich metal 63
arsenical bronze alloys 29, 32, 33, 40, 44, 45, 46, 50, 52, 53, 54, 55, 56, 57, 59, 62, 63, 66, 68, 69, 126
Arsenical Copper Age 4
arsenical copper deposits 29
arsenide ore 59
arsenopyrite ore 29, 55, 77
arsenopyrite-arsenic mineral metallurgy 29
Arslantepe 12, 14, 26, 31, 41, 44, 47, 48, 49, 50, 51, 52, 56, 63, 67, 68, 104, 126, 127
Asia 4, 5, 10, 18, 86
Assyrian merchants 11, 72, 98
Assyrian trading colonies 11, 14, 15, 75
awl 19, 58, 63, 105
awl, bronze 53, 66, 74
awl, copper 1, 20, 21, 24, 31, 52
awl, bone 41
awl, malachite 20
ax 26, 32, 33, 41, 46, 47, 51 n. 8, 65, 67
ax, groundstone 104
ax, clay model 33
ax, bronze 46
ax, copper 64
azurite 20, 24, 59, 77
Aşıklı Höyük 19, 22, 23

Bakır Çukuru 23
Balkans 3, 28, 125
bar ingot 123
battering tool 2
bead 19, 23, 24
bead, azurite 20
bead, cerussite 24
bead, copper 20, 22, 24, 31, 50
bead, galena 24
bead, gold 50
bead, lead 49, 54
bead, malachite 1, 20, 21, 22, 23, 24

bead, silver 50
bead, stone 20, 41
Beersheva culture 104
Belbaşı 18
Beldibi 18
Bereketli Maden 115
beveled-rim bowl 48, 62
Beycesultan 67
Bilgi, Ö. 46
bismuth 52, 55
black burnished ware 87, 101
Black Sea 2, 10, 46, 53, 60, 72, 78, 83, 89, 125
blade 41, 50, 50 n. 5, 52, 54
blade, bronze 50, 61
blade, obsidian 22
Bolivia 9
Bolkardağ (Bulgar Maden) 23, 24, 54, 70, 71, 76, 77, 78, 79, 80, 83, 102
bornite 54
bowl furnace 28, 115, 116, 122
bracelet 46
bracelet, bronze 63
bracelet, copper 22, 32
bracelet, glass 91
bracelet, silver 63 n. 10
bracelet, stone 22
Braidwood, L. 20
Braidwood, R. 7, 20
Britain 11
bronze (*see also* arsenical bronze alloys, tin bronze) 8, 14, 21, 46, 47, 48, 63, 64, 66, 67, 68, 71, 75, 81, 108, 114, 123
Bronze Age 4, 9, 69, 72, 127
Bronze Age finds 85, 91, 92
Bronze Age mining 89, 90, 95, 97
bronze alloying 28, 66, 105
bronze artifacts 46, 47, 48, 63
bronze prills 74
bronze production 66
bronze working 45
bucking tool 87
Bulgar Maden (*see* Bolkardağ)
Bulgaria 89
Burçdere 80
Byzantine 34, 79, 87, 90, 91, 92, 95, 97, 100

calcium 57, 117, 118
calcium carbonate 118, 119, 120
Çamardı 70, 71, 80, 81, 83, 88, 92, 111, 115
Çambel, H. 20
Can Hasan 31, 32
Caneva, I. 64, 65
carbonate ore 19
carnelian 68

cassiterite 71, 73, 74, 80, 81, 88, 90, 91, 93, 100, 106, 112, 113, 114, 115, 117, 119, 120, 121
cassiterite smelting products 114
casting (*see also* lost wax casting) 2, 3, 4, 13, 14, 17, 18, 29, 33, 48, 51, 52, 59, 61, 66, 67, 108, 109, 126
casting ladle 60, 61
Çatal Hüyük 23-25
Caucasus 3, 5, 17, 72, 125
Çayönü 1, 6, 19, 20, 21, 22
Celaller 71, 80, 83, 84, 103, 107, 121, 122
celt 63
celt, stone 20
cerussite 24, 54, 68, 77
chaff-faced ware 48, 49, 58, 62, 87
chalcocite 54, 77
Chalcolithic 4, 6, 11, 12, 14, 15, 17, 25, 27, 28, 29, 30, 31, 34, 36, 40, 41, 42, 45, 47, 55, 56, 57, 58, 60, 61, 62, 63, 79, 87, 88, 89, 101, 104, 126, 127
Chalcolithic, Early 3, 26, 30, 31, 34
Chalcolithic, Late 26, 34, 41, 44, 45, 47, 48, 52, 57, 60, 63, 63 n. 10, 64, 67, 88, 92, 95, 96, 126
chalcopyrite 40, 54, 55, 58, 62, 77
charcoal fuel 2, 38, 42, 56, 57, 59, 78, 87, 88, 92, 94, 120, 122
charcoal production 3, 78
charcoal sample 90, 91, 93
chasing 67
Chatalka 28
Childe, V. 5, 7, 8
chisel 32, 51, 63, 65, 66, 66 n. 6
chisel, bronze 52, 53, 55
chistel, copper 49, 52
chrysocolla ore 52
Cilicia 26, 30, 31, 45, 81, 82, 84, 102, 126
Cilicia Soli Pompeiopolis 50 n. 5
Cilician Gates 76, 85
Cilician plain 80
cinnabar 24, 80
classical 11, 79, 95
cloisonné 14
coarse chaff ware 87
coarse ware 87, 92, 102, 116
Coba/Sakçegözü 31
cobalt 7, 76
cold working 4, 8, 18, 21, 22, 24, 52
complex alloys 56
copper 1, 4, 6, 8, 12, 13, 14, 19, 20, 21, 22, 23, 24, 25, 28, 29, 31, 32, 33, 36, 37, 39, 42, 43, 45, 46, 47, 49, 50, 51, 52, 54, 55, 56, 57, 59, 61, 62, 63, 64, 66, 68, 69, 74, 76, 89, 96, 107, 111, 114, 117, 123, 126
Copper Age 4, 68
copper alloy 14, 54, 74

INDEX

copper alloying 45, 123
copper artifacts 1, 3, 19, 21, 24, 36, 45, 49, 54, 59, 62, 66, 67, 68, 74, 126
copper carbonate 19, 40
copper industry 36
copper metal prills 36, 37, 43, 55
copper metallurgy 38, 112
copper minerals 40
copper nodules 19
copper ore 7, 18, 22, 24, 35, 36, 39, 49, 52, 53, 55, 58, 60, 62, 64, 66, 89, 120, 126
copper oxides 18, 19, 24, 40, 54, 56, 59
copper production 24, 28, 35, 36, 37, 38, 41, 42, 43, 58, 63, 120
copper slag 35, 36, 38, 58, 60, 61, 74
copper smelting industry 126
copper smelting point 57
copper sources 19, 20, 21, 52, 60
copper sulfide 54, 56
copper working 38, 68
copper-arsenic ore 59
copper-arsenic-antimony-lead-nickel alloy 57
copper-based artifacts 32, 46, 69, 75, 115, 126
copper-enriched ore zones 52
copper-iron sulfides and oxides 54
copper-sulfide matte 56
Cornish gold 81
Cornish tin 81
Cornwall 71, 81, 111, 115, 117, 119, 121, 122, 123
covellite 77
Crift Farm 115, 119
crucible 16, 25, 28, 35, 36, 39, 41, 42, 51, 52, 53, 55, 56, 57, 59, 60, 61, 62, 74, 87, 88, 97, 100, 103, 104, 106, 107, 109, 111, 112, 114, 115, 116, 117, 118, 119, 120, 122, 123
crucible melting 24
crucible melting/smelting operations 39
crucible smelting 36, 39, 56, 111, 114, 117, 121
Çukur 62
cuprite 20, 39, 40, 42, 54, 59
cuprous mineral 6
cylinder seal 63
Cyprus 10, 72, 104

dagger 63
dagger, copper/silver 54
dagger, iron with gold handle 69
dark burnished ware 87, 90, 92, 93, 94, 101, 102
dark unburnished ware 92
dark-faced burnished wares 31, 38, 44, 57
delafossit 59

Değirmentepe 12, 13, 26, 27, 28, 30, 31, 32, 33-44, 58, 65, 126
Dolnoslav 28
Dündartepe 50 n. 5

Earl, B. 89, 121
Early Bronze Age 4, 12, 14, 15, 17, 27, 40, 41, 50, 50 n. 5, 51, 53, 56, 63, 67, 68, 69, 70, 71, 75, 79, 81, 83, 84, 87, 88, 90, 90 n. 5, 91, 92, 93, 94, 95, 96, 98, 100, 101, 121, 126, 127
Early Bronze I 41, 46, 48, 51, 60, 61, 63, 74, 95, 96, 101, 108
Early Bronze II 55, 61, 63, 67, 75, 101, 102, 107, 108, 114, 122
Early Bronze III 22, 67, 96, 102, 108, 114
Early Dynastic 14, 50
Elazığ 31, 38, 51, 57, 61
electrum 14, 47, 67, 77
Epipalaeolithic 19
Ergani Maden 19, 20, 21, 60, 62, 64
Erzgebirge 72
Erzurum 51
Esin, U. 21, 22, 23, 33, 34, 36, 37, 61, 62, 64, 69, 75
Eskiyapar 68
Etibank 76
Euphrates 13, 28, 30, 33, 34, 44, 49, 50 n. 5, 51
Europe 5, 6, 11
experimental alloying 25, 32, 64, 65, 66, 126
experimental smelting 112

fahlerz 52, 59
fayalite 59, 61
Fenan 89
figurine 47, 67, 87
figurine, bronze 69, 74
filigree 14, 67
fine monochrome buff-painted wares 30
fine slipped ware 94, 102
fine ware 44, 102
flint-scraped Coba ware 31, 38, 44, 57
forging 18, 29, 33, 69
Fraktin 31
Frangipane, M. 19, 44, 48
Frankfort, H. 5
fruit stands 62
furnace (*see also* natural draft furnace, bowl furnace) 12, 34, 36, 37, 38, 39, 40, 41, 42, 56, 58, 61, 72, 79, 103, 105, 106, 111, 115, 117, 119, 123
furnace-smelting technology 42, 111, 114

Gale, N. 75
galena 24, 28, 54, 62, 68, 77, 78
gallery 88, 91, 94, 96

INDEX

Garstang, J. 64, 65, 66
Garyanın Taşı 79
Gavur Höyük 47, 67
Gawra 43, 46, 58, 65
Gilmore, G. 75
Göbelki 19
Godoy, R. 9
gold 7, 11, 14, 30, 31, 45, 47, 56, 68, 71, 73, 77, 78, 80, 81, 98, 105, 115, 121
gold alloy 14
gold artifacts 67
gold mines 73
Gold Rush 73
gold source 81
gold veins 73
gold plating 69
goldsmiths 13
Göltepe 1, 4, 11, 12, 14-15, 16, 27, 29, 69, 71, 72, 73, 74, 75, 82, 83, 84, 85, 87, 95, 96, 97, 98-109, 111-123, 127
granulation 14, 67
graves 6, 15, 32, 33, 45, 46, 50, 67, 68, 95, 96
ground ore powders 74
groundstone tools 2, 20, 21, 34, 35, 37, 38, 41, 56, 58, 74, 85, 86, 87, 88, 89, 101, 104, 105, 106, 107, 108, 112
Gumelnitsa culture 28
Gümüş slag deposit 79
Gümüşköy 79
Güzelova 51

Haci Nebi Tepe 13, 28
Hacılar 25, 26, 31
Halaf 61, 65
Halaf decoration 64
Halaf-Ubaid 63, 65
Hallam Çemi Tepe 19, 23
hammer 37, 41
hammer, stone 86, 89
hammering 3, 19, 21, 22, 23, 32, 52, 66, 67, 89
hammerstones 2, 37, 69, 90, 91, 93, 95
Harmankaya 33
Hassek Höyük 51 n. 8, 63-64
Hauptmann, H. 52, 57
hearth 34, 35, 36, 37, 38, 39, 42, 51, 58, 63, 65, 92, 93, 103, 105, 106, 107, 120
hematite 61, 69, 73, 74, 76, 80, 81, 91, 96, 98, 104, 112, 113, 114
Hindu Kush 71
hook 58
hook, copper 20, 22
hook, malachite 20
Horoztepe 50 n. 5, 68
horseshoe-shaped installation 36

Ikiztepe 45-46, 50 n. 5, 51, 67

Ilıpınar 45
imported eastern ware 31
Indonesia 9
ingot 15, 63, 105, 127
ingot, bronze 30, 63
ingot, copper 36
ingot, lead 107
ingot, silver 68, 69
ingot, tin 123
intensive surveys 78, 98
intentional alloying 29, 65
Iran 3, 6, 14, 17, 21, 30, 43, 44, 47, 125
Iraq 17, 18
iron 4, 12, 17, 18, 20, 28, 30, 39, 40, 41, 51, 52, 53, 54, 55, 56, 57, 59, 61, 62, 69, 70, 76, 91, 97, 111, 114, 117, 118, 119, 122, 123
Iron Age 4, 9, 34, 57, 79, 94
iron ore 36, 39, 54, 55, 56, 62, 69, 70, 81, 91
iron oxide 54, 120
iron silicates 24
iron-rich flux 62
iron-rich polymetallic ore 56
iron-rich tin ore 72
Israel 6, 42, 47, 89, 104

jewelry 5, 46, 50, 67
jewelry, gold 45
jewery, silver 68
Jordan 89

kalottenförmige vessel 57
Karahöyük 102
Karataş 68
Karaz 47, 51, 67
Keban 31, 51, 62
Keban Dam Salvage Projects 57
Keban mines 55
Kestel 1, 12, 14-5, 16, 29, 55, 71, 72, 73, 74, 75, 80, 81, 82, 83, 85-97, 98, 99, 100, 101, 105, 106, 108, 109, 111, 112, 113, 114, 115, 117, 118, 120, 127
Kestel Surface Survey 85, 99
Khirbet Kerak 51, 51 n. 7
Kısabekir 60
Kish 6, 33
Kızıltarla 21
Konya plain 99
Konya region 23, 102
Korucutepe 51, 69
Koşk Höyük 65
Kültepe (Kanesh) 12
Kunç, Ş. 38, 62
Kur-Araxes 51, 51 n. 7
Kurban Höyük 31
Küre 60
Kuruçay 71, 80, 83, 99

INDEX

Kusura 69, 74
Kütahya 90

lapis lazuli 68
Late Bronze Age 86
Laurion 59
lead 4, 24, 28, 31, 32, 54, 55, 56, 62, 63, 64, 66, 68, 69, 76, 78, 126
lead artifacts 23, 54, 63, 68
lead isotope 24, 45, 59, 63, 64, 76, 127
lead ore 54, 61, 76
lead silicates 57
lead sulfides 30, 68
lead trinkets 33
lead-rich mines 76
lead-zinc mineralization 77
lead-zinc-copper polymetallic Taurus ores 54
Leilan 13
light clay miniature lug ware 95, 102
light ware 44
limonite 36, 77
local ore 15
lost-wax casting 67, 68
low-bronze alloys 28

M.T.A. (Turkish Geological Research and Survey Institute) 71, 76, 81, 88
macehead 32, 46, 61, 63, 69
Madenköy 79
Madenköy slag mound 79
magnetite 59, 76, 113, 114
Mahmatlar 68
malachite 1, 18, 20, 21, 22, 23, 24, 40, 42, 54, 59, 61, 77
Malatya 14, 31, 33, 48, 51, 104
Malaysia 71
manganese 117, 119
manganese oxide 81
Medieval 79, 87, 90, 91, 94, 115, 119
Mediterranean 17, 32, 64, 75, 76, 77, 80, 81, 82, 85, 96, 100, 125
Mellaart, J. 23, 24
Mellink, M. 102
melting 4, 24, 29, 36, 39, 69
melting point 29, 42, 59, 118
Mersin 26, 31, 32, 64-66, 100, 101, 126
Mesolithic 19
Mesopotamia 3, 5, 6, 10, 11, 12, 13, 14, 17, 25, 26, 30, 31, 32, 33, 34, 35, 43, 44, 47, 49, 50, 50 n. 5, 67, 68, 75, 76, 81, 96, 108, 125, 126
metal sources 2, 6, 26, 102
metallic prills 116
metallic ware 87, 102
metallurgical installation 41, 82
meteor 69

micaceous finished ware 92, 94
micaceous slipped ware 94
micaceous unfinished ware 87, 90, 94
micaceous ware 102
Middle Bronze Age 55, 72
Mine Damı 80
miner 3, 9, 12, 73, 78, 83, 88, 91, 111
mineral 1, 11, 13, 19, 23, 24, 29, 39, 40, 51, 55, 66, 73, 74, 77, 86, 90, 91
mineral artifacts 19, 20
mineral composition 80
mineral deposits 15
mineral identification 93
mineral pigments 4, 24
mineral reserves 76
mineral resources 76
mineral samples 91
mineralization 73, 77, 80
mineralogical analysis 77, 91
mold 46, 47, 51, 52, 66, 67, 123, 126
monochrome ware 44
monozite 80
mortar 86, 93, 97, 103, 106, 107, 108
mortar and pestle 19, 93, 104
Muhly, J. 21
Murgul 89, 90

Nahal Mishmar 6, 47
nail, clay 41
nail, wooden 43
native copper 4, 6, 18, 19, 20, 21, 22, 23, 24, 59
native copper artifacts 1, 20, 21
native copper ores 21
native ore 1, 8
natural alloys 1, 66
natural draft furnaces 34, 35, 36, 38, 39, 41, 42, 60, 126
necklace, bronze 46
necklace, copper 19
necklace, silver 107
necklace, stone 22
needle 23 n. 1
needle, bronze 46, 66, 74
needle, copper 66
Neolithic 11, 15, 23-25, 31, 64, 86
Neolithic, Aceramic 1, 3, 6, 17, 18, 19-23
Neolithic Revolution 8
Nevali Çori 19, 28
Nevşehir 104
nickel 28, 32, 46, 52, 53, 54, 55, 62, 63, 64, 69, 70, 126
nickel arsenide 59
nickel-arsenic sulfide ore 54
nickel-rich ore 55
Nissen, H. 45
Niğde 65, 71, 80, 82, 92, 99, 102

Niğde Massif 71, 76, 80, 83, 84, 99, 115
Norşuntepe 27, 28, 31, 32, 41, 51, 57, 58, 59, 60, 61, 64, 67, 126

obsidian 9, 11, 18, 20, 22, 41, 68, 87, 111
obsidian sources 22
obsidian tool industry 22
ochre 20, 36, 37, 69
olivinite 54
orange gritty ware 87, 90, 102
ore 2, 3, 7, 11, 12, 14, 15, 16, 18, 19, 21, 23, 24, 25, 28, 29, 32, 33, 34, 36, 37, 38, 39, 40, 41, 42, 45, 51, 52, 53, 54, 55, 56, 59, 60, 61, 62, 64, 65, 66, 68, 71, 72, 73, 74, 76, 77, 78, 80, 81, 87, 88, 89, 90, 91, 95, 96, 97, 98, 100, 103, 104, 106, 107, 109, 112, 113, 114, 115, 116, 117, 118, 120, 121, 122, 123, 127
ore analyses 88
ore crushing 85, 86, 87, 88, 106
ore deposits 17, 19, 77, 107
ore dressing 77, 86, 87, 88, 90, 93, 99, 100, 104, 112
ore extraction 94, 96
ore powder 122
ore processing 86, 95, 96, 97, 98, 103, 104, 112
ore samples 59
ore selection 56
ore sources 6, 7, 26, 28, 60, 82
ore veins 72, 87, 91
ore-body composition 91
ornament 8, 31, 64, 67
ornament, bronze 48
ornament, copper 18, 19, 23, 24
ornament, lead 24
ornament, leaded copper 46
ornament, malachite 23
ornament, stone 20
orpiment 29
Ottoman 78, 79, 111, 115
oxide ores 54, 59
oxidized ores 77
Özbal, H. 22, 38, 39, 68
Özdoğan, M. 20, 33

Palaeolithic 4
Palestine 22, 51
Palmieri, A. 48, 52
pedestal bowls 57
pegmatite 73, 80, 81
pendant 46, 54
pendant, bronze 46, 54
pendant, copper 24
pendant, lead 68
pendant, malachite 18, 21

pigment 18, 20, 23, 24, 34, 36, 37, 39, 58, 69, 98
pin 24, 46, 63, 64, 65, 66, 66 n. 11, 105
pin, azurite 20
pin, bronze 55, 63, 66, 74
pin, copper 1, 19, 20, 24, 49, 50, 61, 63, 68, 96
pin, gold 50, 69
pin, malachite 20
pin, sliver 50, 54
piroksin 59
pithouse 74, 96, 101, 103, 104, 105, 106, 107, 108, 114, 116, 119, 121, 122
plain simple ware 51, 62, 87, 96, 102
plaque, bronze 50, 53
plaque, iron 69
polymetallic copper-antimony-arsenic oxide ore 58
polymetallic lead ore 57
polymetallic ores 3, 13, 28, 32, 43, 52, 54, 56, 59, 61, 66, 68, 76, 126
polymetallic silver ore 68
polymetallic smelting metallurgy 61
polymetallic sources 74, 83
Pontic 2
Porsuk 79
potassium 117
pre-Classical 79
prills 28, 36, 39, 40, 42, 43, 55, 56, 57, 115, 122, 123
projectile points 41
Puglisi, S. 48
Pulur-Sakyol 47, 51, 67
pyrargyrite 77
pyrite 40, 54, 55, 77, 80
pyrotechnological installation 34, 37, 38
pyrotechnology 20, 31
pyrotine 80

quartz 39, 61, 73, 80, 81, 86, 97, 98, 113, 118
quartz minerals 40

radiocarbon analysis 1, 18, 20, 22, 23, 24, 34, 45, 48, 64, 88, 89, 90, 91, 93, 100, 101
realgar 29, 57, 59
reamer, malachite 20
red burnished ware 44, 49, 87, 92, 101, 102
red-black burnished ware 44, 49, 51, 51 n. 7, 60, 62, 92, 94, 95
refractory materials 79, 97
regional site surveys 30
Renfrew, C. 6
repoussée 67
reserved-slip decoration 60
reserved-slip ware 51, 62
ring 58

INDEX

ring, bronze 46, 105
ring, copper 20, 24, 61
ring, silver 49, 54, 68
riveting 66, 67
Roman 34, 79, 88
Rome 11
Rosenberg, M. 23
rubidium 117
Rudna Glava 89
Russia 84
rutile 80

Sarıtuzla 80, 85, 88
scheelite 80
seals 26, 27, 30, 31, 34, 35, 37, 38, 43, 44, 45, 57, 65, 126
seal impressions 27, 34, 37, 38, 43, 44, 46, 49, 51, 57, 58, 126
sealing practices 43
secondary sulfide ores 32
self-fluxing copper silicates 114
Serçeörenköy 45
Serdaroğlu 33
Sertok, K. 52
Sevin, V. 64, 65
shaft and gallery systems 55, 89, 92, 108
Shanidar Cave 18-19
sheet metal 3, 29, 67, 69
sheet metal, bronze 46, 47
sheet metal, copper 20, 21, 22, 24
sheet metal, gold 47
sickle 46
silicates 59, 119
silicon 118
Silifke 50 n. 5
silver 4, 11, 14, 30, 31, 45, 46, 47, 49, 50, 53, 54, 66, 67, 68, 69, 76, 77, 78, 80, 90, 107, 111, 115
silver alloy 14, 54
silver mining 102, 115
silver ore 46, 54
silver plating 69
silver smelting 111, 115
silver standard 68
silver sulfides 77
silver-copper alloys 68
silver working techniques 68
silvering effect achieved by arsenical segregation 53, 46, 69
simple wheel-made pottery 60
SIMS 116, 118, 119
simug 14
Sızma mine 24
slag 14, 24, 28, 34, 35, 36, 37, 38, 39, 40, 41, 42, 43, 51, 52, 53, 54, 56, 57, 58, 59, 60, 61, 62, 63, 76, 79, 80, 90, 104, 109, 111,
112, 113, 114, 115, 116, 117, 120, 122, 123, 126, 127
smelting 2, 16, 18, 24, 28, 29, 32, 33, 36, 37, 39, 40, 41, 42, 45, 53, 57, 58, 59, 60, 61, 66, 68, 69, 70, 73, 77, 78, 97, 98, 103, 106, 111, 113, 114, 115, 117, 120, 126, 127
smelting activities 97
smelting crucible 39, 74, 114, 126
smelting experiments 16, 52, 56, 59, 109, 116, 121
smelting industry 16
smelting operations 25, 118, 121
smelting oxidized copper ores 61
smelting pits 41
smelting process 16, 40, 42, 61, 63, 106, 111, 117, 120
smelting sites 15, 41, 42, 80, 111
smelting stages 123
smelting technology 15, 119, 121
soldering 67
Solecki, R. 19
spear 50, 52, 53, 54, 63
spear, bronze 46, 50, 53
spear, copper 33, 50
spear, silver 50
speiss 59
sphalerite 77
stamp seal 34, 41, 43, 58, 63
stamp seal production 43
stannite 71, 77
statuary 67
Stòs-Gale, Z. 75
strontium 117
sulfide ore 28, 35, 39, 40, 45, 53, 54, 55, 56, 59, 61, 62, 63, 77, 126
Sulu Mağra 80
Sulucadere 77
survey 15, 30, 33, 70, 71, 76, 78, 79, 81, 82, 83, 85, 98, 99, 100, 102, 122
survey methods 82
survey station 90
Susa 43, 50
Susiana plain 30
sword 46, 52, 53
sword, bronze 46, 50, 53
sword, silver 54
Syria 3, 11, 13, 14, 17, 23, 26, 30, 31, 34, 44, 47, 50 n. 5, 51, 67, 75, 76, 81, 84, 96, 102, 108, 125
Syrian bottles 95, 96
Syrian metallic ware 88, 95, 96, 102
Syrian ware 60
Syro-Anatolia 31
Syro-Mesopotamia 11, 15, 27, 48, 49, 62, 126
Syro-Mesopotamian exchange pattern 125

Syro-Palestine 10

tablet/ancient texts 9, 11, 12, 27, 30, 67, 72, 74
Talmessi 59
Tarsus 50 n. 5, 51, 64, 67, 69, 75, 77, 80, 100, 101, 102, 107
Taurus 1, 2, 4, 11, 23, 24, 29, 31, 55, 60, 62 n. 9, 66, 70, 78, 81, 82, 83, 84, 102, 108, 125, 127
Taurus 2B 60
Taurus, cental 71, 72, 75, 76, 80, 84, 85, 102, 127
Taurus, eastern 127
Taurus mineralization 23
Tell al-Judaidah 13, 27, 31, 47, 50 n. 5, 66 n. 11, 74
Tell Brak 13, 50 n. 5
Tell Kurdu 31, 32, 69
Tello 33, 43, 50
tennantite 54
Tepe Giyan 43
Tepe Sialk 43
Tepecik 27, 31, 41, 51, 62, 63, 70, 126
ternary bronze 32, 66, 126
textile 11, 12, 24, 41, 75
Thrace 17
Tigris 13, 30, 44
Tigris-Euphrates 30, 31, 45, 63
Timna 36, 42, 89
tin 4, 11, 15, 23, 29, 32, 45, 47, 54, 55, 63, 64, 66, 67, 68, 69, 71, 72, 73, 74, 75, 76, 77, 80, 81, 83, 84, 87, 88, 89, 91, 97, 98, 100, 104, 105, 106, 107, 111, 113, 114, 115, 116, 117, 118, 119, 120, 121, 122, 123, 126, 127
tin alloying 69
tin bronze 23 n. 1, 29, 47, 65, 67, 71, 72, 73, 74, 75, 114, 123, 126
tin bronze alloying 74, 114
tin industry 15, 112
tin mineralization 72, 80, 88
tin mining 15, 80, 91
tin ore 9, 71, 73, 77, 80, 109, 118, 120, 121
tin metal prills 115, 119, 120, 122, 123
tin production 16, 27, 72, 101, 104, 108, 111, 112, 115, 117, 118, 121, 127
tin silicates 113
tin slag 72, 111, 115, 119
tin sources 71, 72, 75, 81
tin-gold anomaly zone 80
tin-rich ore 73, 74, 106, 112, 119
Tirebolu 60
titanite 80
titanium 117, 118, 119
Tokat-Erbaa 90
token 27, 30, 43

tool 1, 5, 6, 8, 9, 12, 13, 26, 27, 31, 32, 41, 42, 46, 47, 64, 65, 66
tool, bone 24, 41
tool, bronze 41, 46, 48
tool, copper 12, 19, 21
tool, stone ore processing 8, 73, 86, 87, 89, 90, 91, 93, 95, 100, 106, 112
tool, weaving 41
Transcaucasian 51, 60, 62
Transcaucasian ware (*see* red-black burnished ware)
transhumance 84
Troad 72
Troy 52, 67, 68, 69
tube 24
Tülintepe 27, 31, 41, 61-62, 69, 126
Turkish Geological Survey 79, 82, 88
tuyere 28, 42

Ubaid 11, 12, 15, 25, 26, 27, 30, 31, 32, 33, 34, 36, 44, 50, 57, 61, 62, 65, 66, 69, 125
Ubaid painted ware 31
Ubaid-related cultural elements 12, 13, 31, 34, 44, 58, 60
Ubaid-related wares 31, 44, 57, 62, 65, 66
Upper Palaeolithic 18
Ur 6, 33, 50, 67
Uruk 13, 15, 25, 27, 28, 44, 45, 48, 49, 50, 58, 62, 63, 64, 66, 96, 125
Uruk-related 11, 12, 13, 14, 44, 60, 61, 62, 63
Uruk-related ware 45, 49, 62, 62 n. 9

Varna 6
vein 25, 73, 74, 76, 77, 80, 81, 88, 89, 91, 114

weapon 1, 5, 6, 8, 9, 12, 13, 26, 27, 31, 32, 65, 66, 67
weapon, bronze 46, 47, 53, 63
wustite 59

Yugoslavia 72, 89

Zagros 18
Zawi Chemi Shanidar 19
Zeytindağ 55
zinc 4, 28, 29, 32, 52, 56, 61, 62, 68, 76, 107, 126
zinc ore 77

Table 1: *Trace Element Analyses of Bolkardağ Ores and Slag Samples (AAS)*
(All samples from sites B 8, B 11, B 31, B 34, B 30, B 37, B 6, B 32)

a: Trace Element Distribution of Iron Oxide Rich Placer Ores

Element	Minimum concentration	Maximum concentration	Average concentration*
Au	0.111 ppm	62.64 ppm	8.86 ppm
Ag	01605.0 ppm	357.2 ppm	
Sn	0	1170.0 ppm	220.0 ppm
Pb	0.03 %	26.32 %	7.09 %
Zn	0.04 %	15.36 %	5.92 %
As	0	8.25 %	3.21 %
Sb	0	0.49 %	0.06 %
Ni	0	0.17 %	0.05 %
Co	0	0.04 %	0.01 %
Mn	0.01 %	10.51 %	1.72 %
Cd	0	0.43 %	0.09 %
Cu	0.03 %	1.57 %	0.30 %
Fe	8.32 %	49.65 %	38.12 %

* Total of 39 different ores

b: Elemental Distribution of Bolkardağ Galena & Sphalerite Ores

Element	Galena Ores*	Sphalerite Ores*
Au	12.21 ppm	7.897 ppm
Ag	527.9 ppm	280.2 ppm
Sn	600.0 ppm	300.0 ppm
Pb	21.32 %	5.27 %
Zn	8.01 %	18.55 %
As	0.71 %	0.73 %
Sb	0.15 %	0.10 %
Ni	tr	tr
Co	tr	tr
Mn	1.30 %	1.53 %
Cd	0.20 %	0.30 %
Cu	0.31 %	0.32 %
Fe	14.03 %	10.73 %

* Total of 10 different ore samples

c: Elemental Distribution of Bolkardağ Slag Samples

Element	Group 1	Group 2	Group 3	Group 4
Au	3.95 ppm	1.79 ppm	6.320 ppm	14.53 ppm
Ag	476.0 ppm	183.0 ppm	212.3 ppm	157.6 ppm
Sn	730.0 ppm	300.0 ppm	640.0 ppm	450.0 ppm
Pb	10.37 ppm	9.26 %	3.32 %	4.83 %
Zn	0.68 %	0.16 %	0.46 %	0.86 %
As	8.74 %	2.36 %	5.93 %	4.01 %
Sb	0.75 %	0.26 %	0.39 %	0.24 %
Ni	0.13 %	0.12 %	0.03 %	0.04 %
Co	0.02 %	0.01 %	tr	tr
Mn	0.11 %	0.04 %	0.14 %	1.14 %
Cd	0.01 %	0	tr	tr
Cu	0.66 %	0.34 %	0.46 %	0.40 %
Fe	31.10 %	30.24 %	34.48 %	33.18 %

Table 2: *Çamardı Sites*
(Map references from 1: 25, 000 map)

Site number	Name	Dates	Map coordinates	Location	Size (ha.)
Ç 1	Sulumağara Mine	EBA	Kozan M33 b3	4.5 km upstream from Kestel	
Ç 2	Kestel Mine complex / Work stations	EBA, Byzantine	Kozan M33 b3	1878 m alt. 2 km West Celaller	1.5 km
Ç 3	Ören Tepe	Halaf-Medieval	Kayseri L33 c4	1409 m alt. 4.5 km East Ovacık Köy	3.0
Ç 4	Göremsen I	Medieval	Kozan M33		0.5
Ç 5	Göremsen II	Medieval	Kozan M33		0.5
Ç 6	Terlik Tepe	Neolithic, Chal, EBA	Kozan M34 a4	100 m SSW Çukurbağ school	0.36
Ç 7	Eğer Tepe	MBA, Iron, Classical, Medieval	Kozan M34 a4	1 km SSW Demirkazık	2.0
Ç 8	Cimbar	MBA, Iron	Kozan M34 a4	500 m N. Demirkazık	0.36
Ç 9	Çallıyurt	Ottoman	Kozan M33 b3	Opposite Kestel Mine	0.09
Ç 10	Ildirözü	EBA, Roman, Medieval	Kozan M33 c2	4 km S. Celaller	2.25
Ç 11	Göktaş	Classical Medieval	Kozan M33 c2	1 km SSE Burç Köy	1.4
Ç 12	Ağaca Mevkii	Medieval	Kozan M33 c2	1 km SSW Burç Köy	6.2
Ç 13	Göltepe	Chal/EBA	Kozan M33 b2	2 km S. Kestel Mine	10.0
Ç 14	Üçkapılı	Medieval	Kozan M33 b2	Üçkapılı Köy	1.6
Ç 15	Gümük	Chalcolithic, EBA	Kozan M33 b2	4 km N. Üçkapılı Köy	0.8
Ç 16	Kaletaşı	Classical	Kozan M33 b2	2 km N. Ören	0.36
Ç 17	Ziyaret Tepe	?	Kozan M33 b3	5 km W. Celaller	1.0
Ç 18	Devetaşı Tepe	?	Kozan M33 b4	1 km NW	1.)

Table 2: *Çamardı Sites (cont'd)*

Ç 19	Karacaören	Medieval	Kozan M33 b4	5 km E. Kılavuz Köy	?
Ç 20	Çakıl Tepe	EBA	Kozan M33 b4	5 km W. Karacaören	9.0
Ç 21	Tazıyoran Kayası	EBA, Iron, Medieval	Kozan M33 b4	1 km SE Kılavuz	11.5
Ç 22	Höyük Tepe	Chalcolithic EBA	Kozan M33 b4	Kılavuz	7.0
Ç 23	Kapı Taşı Mevkii	?	Kozan M33 b4	4 km E. Kılavuz	0.7
Ç 24	Çardaklı	Classical Medieval	Kozan M34 a4	1 km N. Elekgölü	100
Ç 25	Kuruburun Tepe	EBA, Iron, Classical	Kozan M34 a4	6 km S. Çamardı	8.0
Ç 26	Boztepe	Neolithic, Chalcolithic	Kozan M33 b3	1 km E. Mahmatlı	0.6
Ç 27	Karatepeler	EBA	Kozan M33 b3	1 km N. Çardacık	4.5
Ç 28	Kale Tepe	EBA, Iron Medieval	Kozan M33 b3	3 km W. Göltepe	10.0
Ç 29	Karamıhlı	Medieval	Kozan M33 b4	2 km E. Çiftlik	?
Ç 30	Küçük Çanak Tepe	Classical Medieval	Kozan M33 b4	2 km NE Çiftlik	0.37
Ç 31	Büyük Çanak Tepe	Classical	Kozan M33 b4	8 km W. Celaller	7.0
Ç 32	Çevlik Tepe	EBA, Iron, Med	Kozan M33b2	1 km. N. Ören	3
Ç 33	Üçkapılı	Classical	Kozan M33 b2	Üçkapılı Köy	0.03

Table 3: *Kestel Radiocarbon Dates*

Findplace, laboratory number	Sample Description	C14 Date B.P. Date B.C. (Libby) Calibrated B.C. 1 σ range Calibrated B.C. 2 σ range
Kestel 1 Bryan Earl New Zealand/Oxfordshire	Fireset charcoal inside Kestel Mine	- 2858-2468 B.C. 2874-2362 B.C.
Kestel S46 BM-2879	Firehole cavity inside Kestel Mine	4090 ± 60 B.P. 2140 ± 60 B.C. 2865-2810 B.C. 2880-2495 B.C.
Kestel S33 BM-2881	Lowest level, trench 4, mine 1 Kestel	4690 ± 100 B.P. 2740 ± 100 B.C. 3625 - 3360 B.C. 3700-3300 B.C.
Kestel S2 BM-2880	Large chamber, Kestel mine	220 ± 45 B.P. - 1640 A.D.-modern 1915 A.D.-modern
Kestel S24 BM-2882	Opencast working, Kestel slope	1210 ± 50 B.P. - 725-890 A.D. 915-945 A.D.
Kestel I-15,227	Kestel mine, sounding S.2, -68 cm. depth	3980 ± 100 B.P. 2030 ± 100 B.C. 2870-2200 B.C.
Kestel AMS AA-3373	Kestel mine, sounding S.2 -30 cm. depth	1570 ± 60 B.P. 620 ± 60 A.D. 422-547 A.D. 347-609 A.D.

Table 3: *Kestel Radiocarbon Dates (cont'd)*

Kestel AMS AA-3374	Kestel mine, sounding S.2 -68 cm. depth	4020 ± 80 B.P. 2070 ± 80 B.C. 2855-2466 B.C. 2874-2350 B.C.
Kestel AMS AA-3375	Kestel mine, sounding S.2 -60 cm. depth	3805 ± 70 1955 ± 70 2473-2297 B.C. 2576-2147 B.C.
Kestel AMS AA-3376	Kestel mine, sounding S.2 -93 cm. depth	3830 ± 65 1880 ± 65 2456-2147 B.C. 2469-2133 B.C.

Table 4: *Göltepe Radiocarbon Dates*

Main Registry Number (MRN), findplace laboratory	Sample Description	C14 Date B.P. Date B.C. (Libby) Calibrated B.C.1 s range Calibrated B.C. 2 s range
MRN 391 Göltepe A06-0100-011 Beta 42649	Charcoal from hearth (011) on pithouse floor	5240 ± 250 B.P. 3290 ± 250 B.C. 4350-3780 B.C. 4661-3518 B.C.
MRN 296 Göltepe B01-0126-001 Beta 42648 ETH 7669 AMS	Charcoal from crucible middens near circuit wall	4120 ± 60 B.P. 2170 ± 60 B.C. 2875-2587 B.C. -
MRN 1400 Göltepe A24-0343-006 Beta 42650	Charcoal from floor of pithouse 006	4070 ± 60 B.P. 2120 ± 70 B.C. 2863-2812 B.C. 2876-2801 B.C.
MRN 1785 Göltepe A23-0100-007 Beta 42651	Charcoal from burnt debris 20 cm above floor	3790 ± 80 B.P. 1840 ± 80 B.C. 2451-2050 B.C. 2470-1985 B.C.
MRN 2338 Göltepe A22-0300-008 Beta 75607	Ashy layer on floor of pithouse	4020 ± 70 B.P. 2070 ± 70 B.C. 2595-2460. B.C. 2695-2335 B.C.
MRN 2540 Göltepe E63-0100-013 Beta 75609	Pit house fill	3840 ± 70 B.P. 1890 ± 70 B.C. 2440-2175 B.C. 2090-2040 B.C.

Table 4: *Göltepe Radiocarbon Dates (cont'd)*

MRN 2585 Göltepe A23-0900-007 Beta 75610	Pit house fill	3830 ± 60 B.P. 1880 ± 60 2350-2175 B.C. 2080-2050 B.C.
MRN 3156 Göltepe E63-0100-026 Beta 75613	Pit house 006 collapse fill	3910 ± 90 B.P. 1960 ± 90 B.C. 2485-2270 B.C. 2595-2130 B.C.
MRN 599 Göltepe A26-0100-007 Dendrochronology	Pit fill	- - 1978 ± 37 B.C.
MRN 5607 Göltepe D67-0200-017 Beta 75614 AMS	Pithouse fill/ wall phase	4150 ± 50 B.P. - 2885-2575 B.C. 2875-2790 B.C. and 2780-2605 B.C.
MRN 2811 Göltepe D67-0200-016 Beta 075612	Pithouse fill south room/ wall phase	3800 ± 100 B.P. - 2485-1935 B.C. 2400-2110 B.C. and 2090-2040 B.C.
MRN 5665 Göltepe B05-1100-019 Beta-105168	Pithouse B06, lowest pit fill in floor 009	4150 ± 70 B.P. - 2900-2490 B.C. 2880-2590 B.C.

Table 4: *Göltepe Radiocarbon Dates (cont'd)*

MRN 3214 Göltepe C02-0100-006 Beta-104994	Floor, uppermost level	2720 ± 60 B.P. - 990-795 B.C. 910-815 B.C.
MRN 2383 Göltepe B05-1100-009 Bet1-075608	Floor	3890 ± 50 B.P. - 2480-2195 B.C. 2460-2290 B.C.
MRN 2682, 5717 Göltepe B06-0300-012 Beta-075611	Bench	4140 ± 50 B.P. - 2885-2570 B.C. 2870-2795 B.C. and 2770-2595 B.C.
MRN 2223 Göltepe E69-0100-009 Beta 75606	Midden with metallurgical debris	3980 ± 70 B.P. 2030 ± 70 B.C. 2575-2440 B.C. 2620-2290 B.C.
MRN 2079 Göltepe C02-0200-015 Beta 75605	Ashy layer on floor of pithouse, possibly phase 2	3920 ± 60 B.P. 1970 ± 60 B.C. 2475-2310 B.C. 2570-2205 B.C.
MRN 5718 Göltepe B06-0500-006 Beta 104993	Pit and fill	3940 ± 50 B.P. 1990 ± 60 B.C. 2480-2350 B.C. 2570-2290 B.C.

Table 5: *Atomic absorption analysis of metal objects from Göltepe*

Sample	MRN#	Au*	Ag	Sn	Pb	As	Sb	Ni	Zn	Co	Fe	Cu	Bi
Metal Piece	4570	13.4	nd	0.1	0.17	nd	0.01	nd	nd	nd	0.06	90.4	nd
Copper Bracelet	2412	-	0.02	12.2	0.01	0.03	0.06	0.05	0.04	nd	0.08	89.3	0.02
Bronze Pin	1893	-	0.06	11.1	0.24	0.06	0.15	nd	0.07	0.01	0.16	88.0	nd
Copper Spiral	4692	52.1	0.03	11.7	0.81	0.04	0.41	0.05	0.03	nd	0.35	85.3	0.01
Copper Spiral	4692a	38.9	0.01	12.3	0.26	0.08	0.37	0.07	0.02	0.01	0	81.9	0.01
Copper Tube	4672	1.23	0.01	12.0	0.15	0.03	0.02	0.26	nd	nd	0.62	80.0	0.01
Hair Ring	4137	27.6	0.03	4.75	0.05	0.02	0	0.02	0.02	nd	0.29	78.4	0.01
Bronze Needle	3110	-	0.12	1.49	0.89	1.21	0.57	0.02	0.09	0.01	1.08	71.3	nd
Silver Necklace	3993	-	91.2	1.11	0.01	nd	0.25	0.03	1.16	nd	0.41	2.18	0.18
Lead Piece	2604	-	0.11	nd	98.2	0.01	0.61	0.01	nd	0.01	0.09	nd	nd

*Gold results are in ppm; all other results are in percent; nd = not detected.

Table 6: *Comparison of the elemental analysis of hematite samples*

Element (ppm)	Kestel Hematite		Göltepe Hematite		Powdered Material		Magnetic Material	
Au	1.31	(20)	0.55	(6)	0.65	(12)	0.52	(10)
Sn	647	(34)	2080	(15)	4464	(60)	2571	(12)
Ag	8.35	(20)	12.0	(6)	7.95	(21)	51	(10)
Pb	285	(20)	150	(6)	481	(21)	420	(10)
As	1395	(20)	750	(6)	1395	(18)*	34	(10)
Sb	600	(20)	450	(6)	650	(12)	1135	(10)
Ni	55	(20)	83	(6)	124	(21)	180	(10)
Zn	135	(20)	100	(6)	257	(21)	106	(10)
Co	0	(20)	0	(6)	0	(21)	0	(10)
Cu	85	(20)	200	(6)	386	(21)	7160	(10)
Fe (%)	31.9	(20)	39.9	(6)	29.4	(21)	43.4	(10)

Iron results are in percent; all others are in ppm; all results represent averages; numbers in the parentheses represent the number of samples analyzed.

*Three samples, MRN 3841, MRN 4774, and MRN 3842, had unusually high arsenic values of 5.58%, 6.02%, and 5.54%, respectively, and are not included in the average.

Table 7: *Atomic Absorption analysis of Göltepe powdered samples*

Sample #: Analysis #:	MRN 3841 (92/300)	MRN 3842 (92/314)	MRN 4774 (92/309)	MRN 2611 (93/315)	MRN 3737 (93/314)
Sn	1.12	1.40	0.72	0.45	0.66
As	5.58	5.54	6.02	0	0.01
Pb	0.01	0.01	0.02	0.03	0.03
Sb	0.12	0.01	0.01	0	0.10
Ag	-	-	-	0	0
Ni	0.01	-	0.01	0.02	0.02
Bi	0.05	0.04	0.04	0.05	0.05
Au (ppm)	0.92	1.34	0.46	0.61	1.08
Zn	0.02	0.01	0.01	0.01	0.01
Cu	0	0	0	0.01	0.01
Fe	15.2	19.4	28.2	35.0	31.5
Vanning Sn (Estimate)	0.25%	0.5%	No Head	No Head	Slight Head

Gold results are in ppm; all others are in percent.

Table 8a *General sample information*

Sample no.	Original reference no.	Site location	Site context
1	MRN 2298	E70-0100-003	Midden
2	MRN 2836	A15-0100-005	Floor of pit house
3	MRN 3032	B05-1000-074	Secondary fill deposit
4	MRN 3697	B06-0300-014	Secondary fill deposit, next to hearth in pit house
5	MRN 3738	E63-0400-001	Secondary fill deposit
6	MRN 3830	A14-1000-003	Pit house floor inside ceramic cup
7	MRN 3834	A14-1000-003	Same pit house floor as MRN 3830, inside ceramic cup
8	MRN 3858	A14-0700-003	Same pit house as MRN 3830 and MRN 3834, fill over floor
9	MRN 4573	A02-06-034	Above floor of pit house

Table 8b *Elemental concentrations (in wt% or ppm) measured by X-ray fluorescence*

Element	Unit	Sample 1	Sample 2	Sample 3	Sample 4	Sample 5	Sample 6	Sample 7	Sample 8	Sample 9
K	%	0.61	0.14	1.20	1.40	0.42	0.46	0.43	0.63	0.50
Ca	%	9.34	2.55	9.70	7.30	4.90	7.30	15.80	7.60	4.20
Ti	%	0.35	0.34	0.28	0.69	0.062	0.10	0.07	0.11	0.07
Fe	%	24.31	54.40	16.60	33.10	28.90	41.00	6.90	34.30	21.80
As	%	0.43	0.09	0.05	0.22	0.15	0.33	0.66	0.10	0.08
Sn	%	0.28	0.85	0.34	0.64	0.70	1.18	0.43	0.85	2.93
V	ppm	173	155	–	303	–	118	–	125	–
Cr	ppm	236	263	141	1300	–	–	–	–	–
Mn	ppm	699	505	570	902	370	603	817	615	200
Ni	ppm	43	–	55	43	35	47	22	52	18
Cu	ppm	189	42	45	209	97	99	29	86	39
Zn	ppm	88	49	77	89	56	57	52	60	58
Br	ppm	23		–	13	13	17	24	–	–
Rb	ppm	21	13	57	41	14	22	–	22	–
Sr	ppm	148	50	268	275	83	185	196	152	72
Zr	ppm	597	1800	174	938	18	39	36	48	–
Bi	ppm	149	115	29	75	63	122	46	70	162

Table 9 *Average composition of the tin-containing particle types in the nine powder samples*

Average abundance (%)	Group			
	Tin oxides	Tin silicates	Fe-Sn rich	Sn-Fe rich
Al_2O_3	0.1 ± 0.1	3 ± 2	0.3 ± 0.2	0.5 ± 0.7
CaO	–	9 ± 18	8 ± 7	–
Fe_2O_3	5 ± 3	10 ± 5	53 ± 7	25 ± 6
SiO_2	4 ± 2	43 ± 20	5 ± 5	5 ± 5
SnO_2	91 ± 4	32 ± 15	33 ± 15	69 ± 5

Table 10 XPS results

Sample (1)	Sum of all Sn peaks (2)	SnO (3)	Sn (4)	Percentage of metallic Sn (5)
1	6623	6623	–	–
2	43948	43558	390	0.89
3	7886	7886	–	–
4	22606	21361	1245	5.51
5	12466	12466	–	–
6	11185	11185	–	–
7	1667	1667	–	–
8	1120	1120	–	–
9	35733	33968	1765	4.94

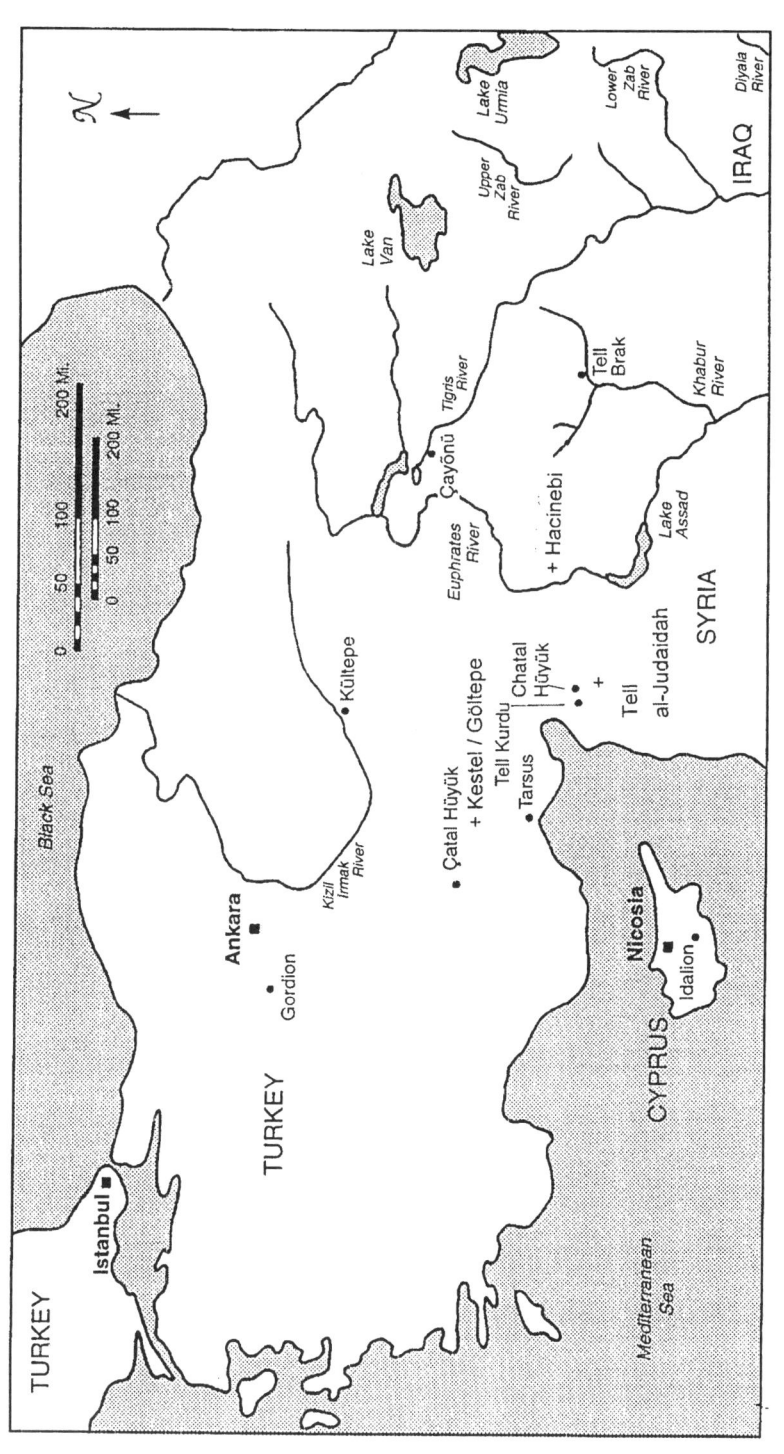

Fig. 1: Map of Turkey

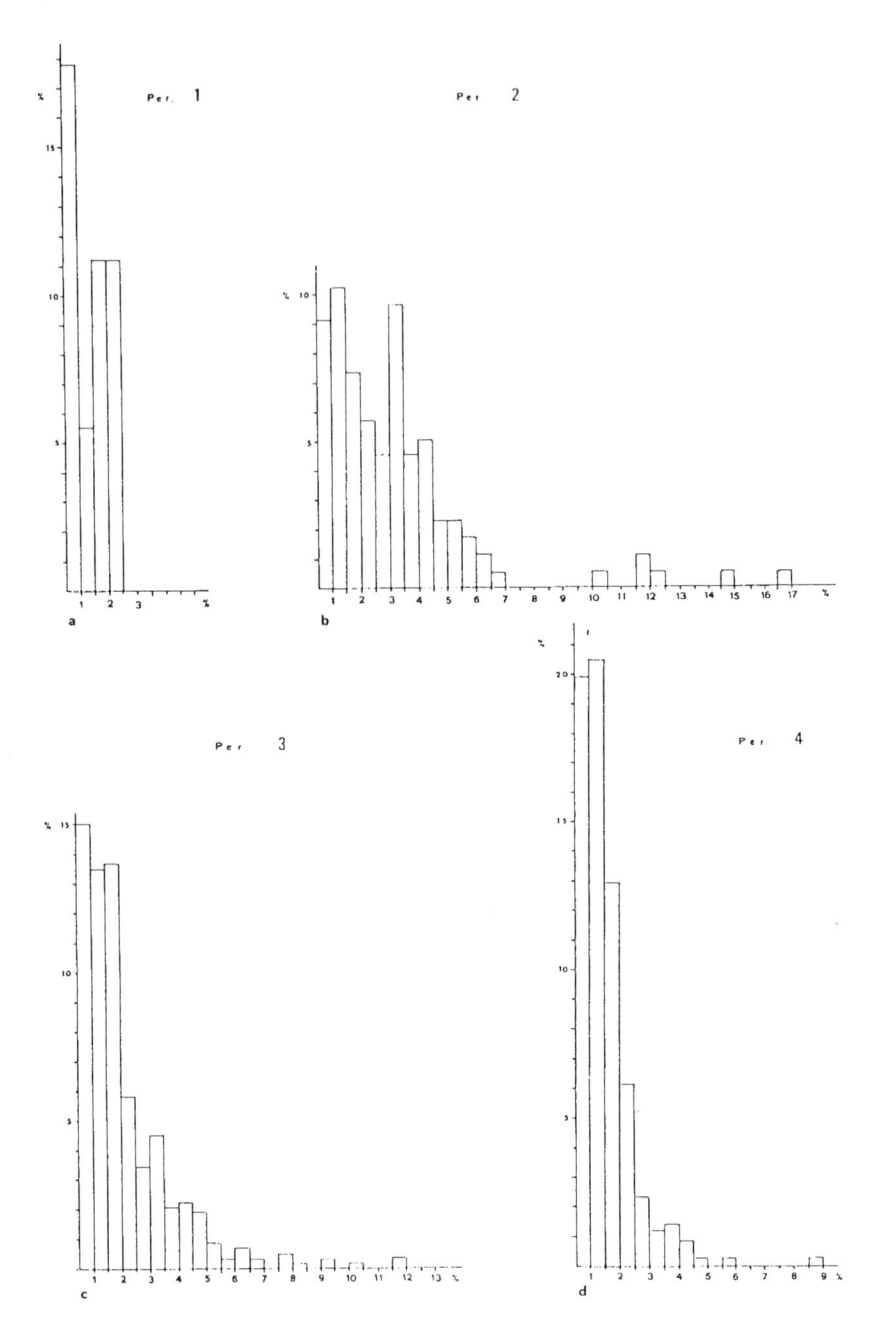

Fig. 2: a: Histogram of arsenic content in Near Eastern copper and bronze objects; b: histogram of arsenic content in tinless copper objects. Period 2: late 4th-early 3rd; Period 3: late 3rd; Period 4: Middle Bronze Age. From Caneva, Frangipane, and Palmieri 1985: 128, Fig. 6

Metallurgy At Değirmentepe

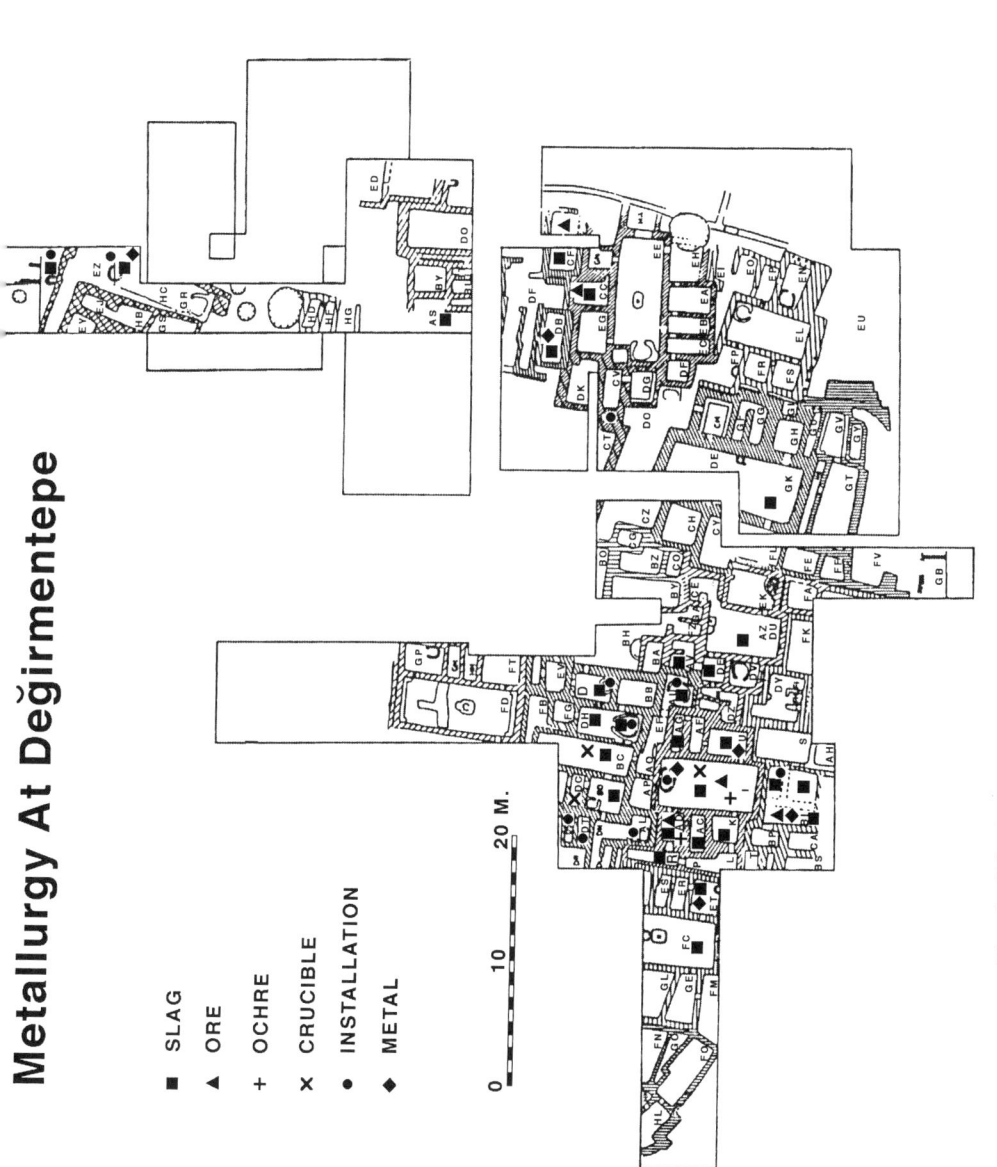

- ■ SLAG
- ▲ ORE
- + OCHRE
- × CRUCIBLE
- ● INSTALLATION
- ◆ METAL

Fig. 3: Distribution of metallurgical debris from Değirmentepe. After Esin 1989

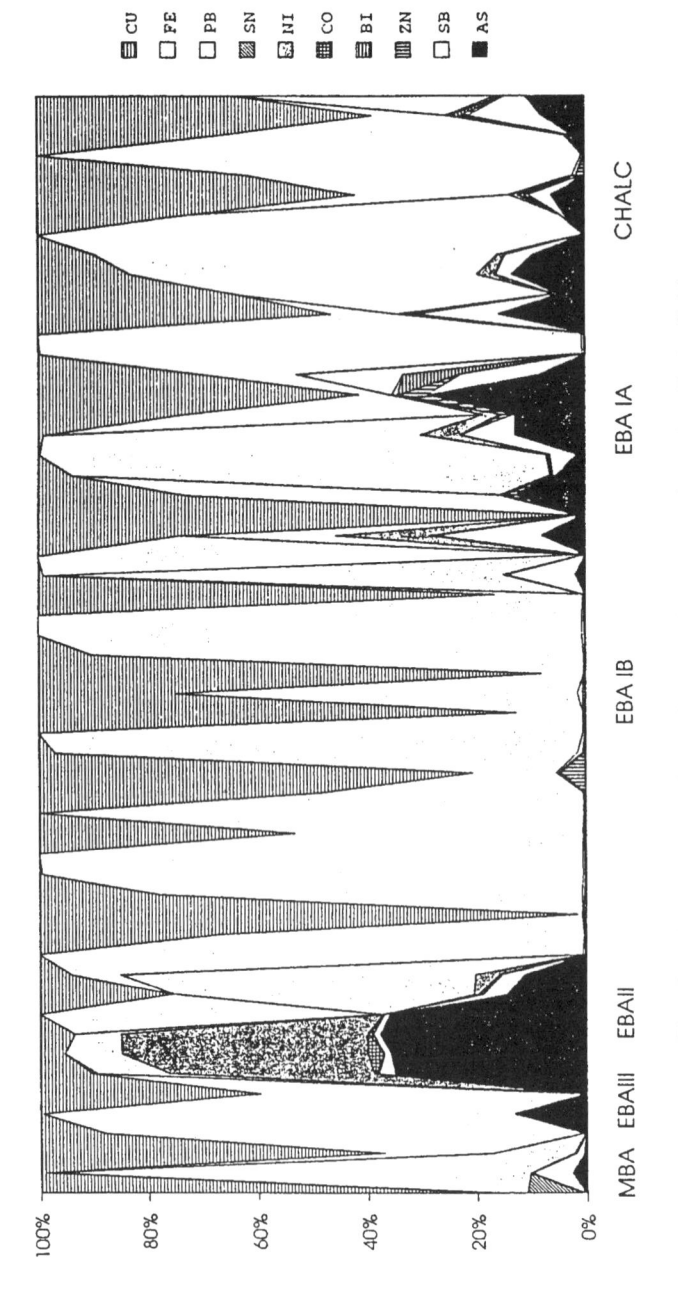

Fig. 5: Metal content of ores found at Arslantepe from the Chalcolithic to the MBA. Note the absence of As, Sb, Ni and Pb in the samples from the EBA IB! (elements listed recalculated to 100%). From Palmieri, Hauptmann, Hess, and Sertok 1996: Fig. 1.

Fig. 4a: A bimodal distribution is indicated for the differences between arsenic contents of swords versus spears. From Caneva, Frangipane, and Palmieri 1985: 117

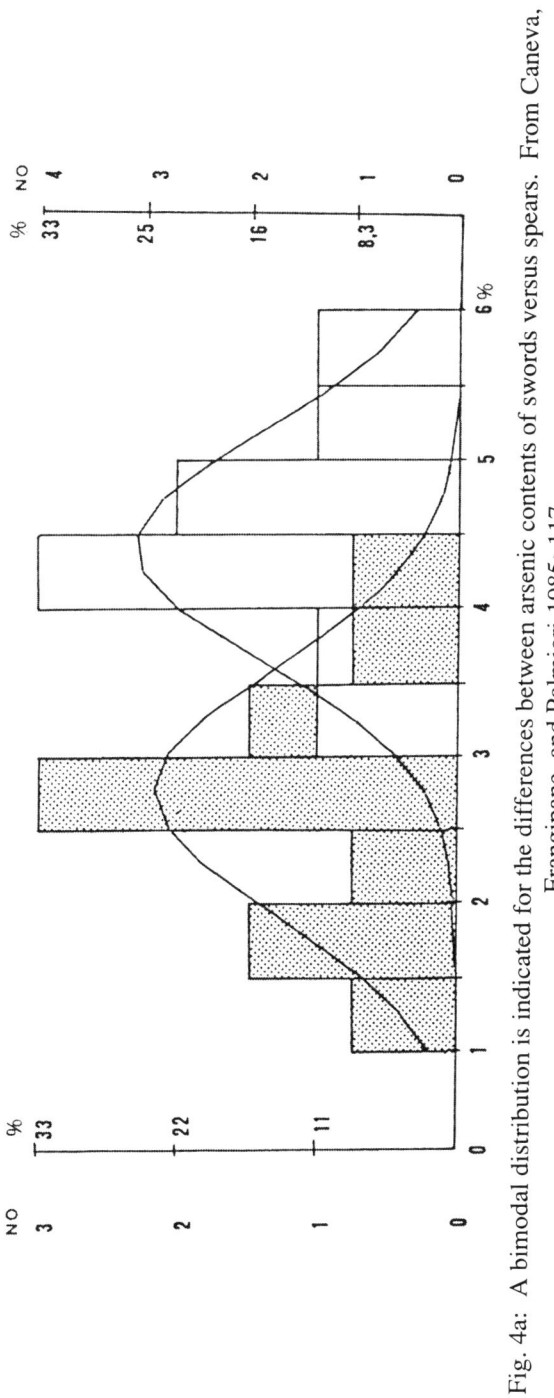

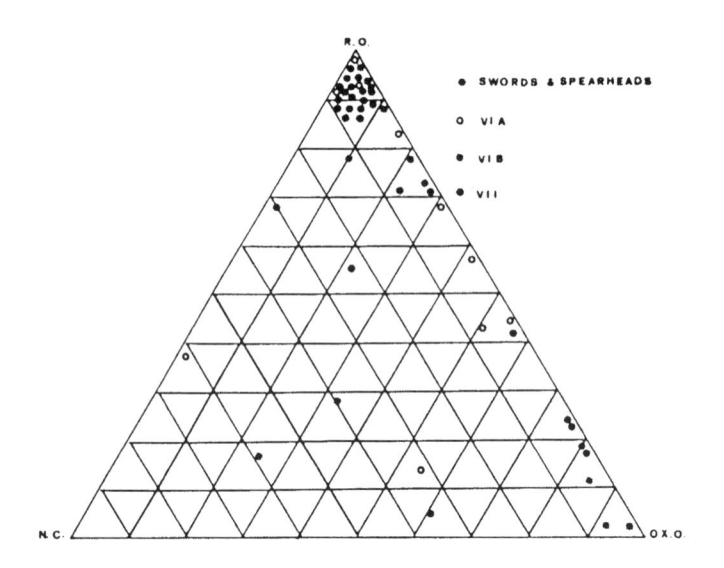

Fig. 4b: A ternary diagram of the trace elements in the artifacts suggests that most were derived to a lesser extent from oxides and sulfides. From Caneva and Palmieri 1983: 643.

Fig. 6: Topographical map of the Bolkardağ region. B1. Yayvantepe; B2. Mahmutsekisi; B3. Aktaş Tepe; B4. BULGARMADEN Inscription; B5. Yediharmantepe-Kaltakbeleni workshops; B6. Sulucadere Mines; B7. Madenköy; B8. Yeşelli, Küçük Toyislam, K.H. Mines; B9. Çingenetepe; B10. Karyayla Tepe; B11. Bakırtepe Mines; B12. Kalkankaya Mevkii; B13. Gümüş Mevkii; B14. Gümüş Yayla; B15. Ilhan Yayla; B16. Gümüsköy; B17. Eğercinin Döleği Mevkii; B18. Çatal Ağzı-Haram Boğazı-Mezarın Tepe cemetaries; B19 Katırgedigi site; B20. Pancarcı Kale; B21. Geyik Pınar Kale; B22. Kocanın Çamı Grave; B23. Tabaklı Kale; B24. Karagümüş Mevkii; B25. Tavşanın Yeri Mevkii; B26. Garyanın Taşı Tepe; B27 Solağın Yeri Mevkii; B28. Tekne Çukur Tepe; B29. Göğceli Mevkii; B30. C. and D. Galleries, D-5 Mine, Orta Mine; B31. Öküzgönü Mines; B32. Kızıltepe, Bıstırgan, and Gavurlar Yurdu Mines; B33. Katırgediği Kuzeyi Mines; B34. Korucuk and Selamsızlar Mines; B35. Sulucadere site; B36. Darboğaz Mevkii; B37. Sulu Mağara and Kara Mağara Mines.

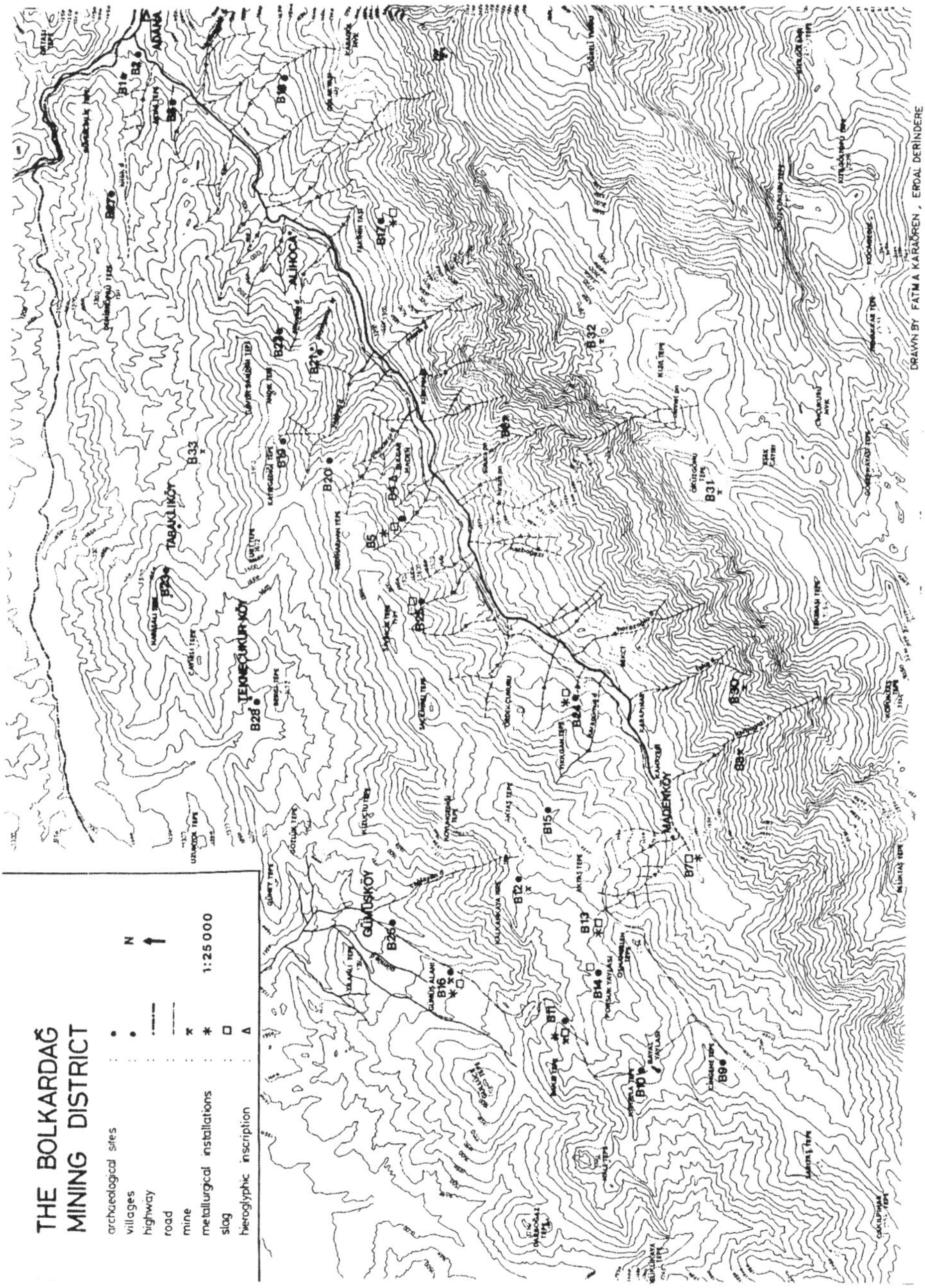
THE BOLKARDAĞ
MINING DISTRICT
archaeological sites
villages
highway
road
mine
metallurgical installations
slag
hieroglyphic inscription
N
1:25 000
ADANA
ALİHOCA
TABAKLIKÖY
TEKNECUKUR-KÖY
GÜMÜŞKÖY
MADENKÖY
DRAWN BY FATMA KARAÖREN , ERDAL DERİNDERE

190 FIGURES

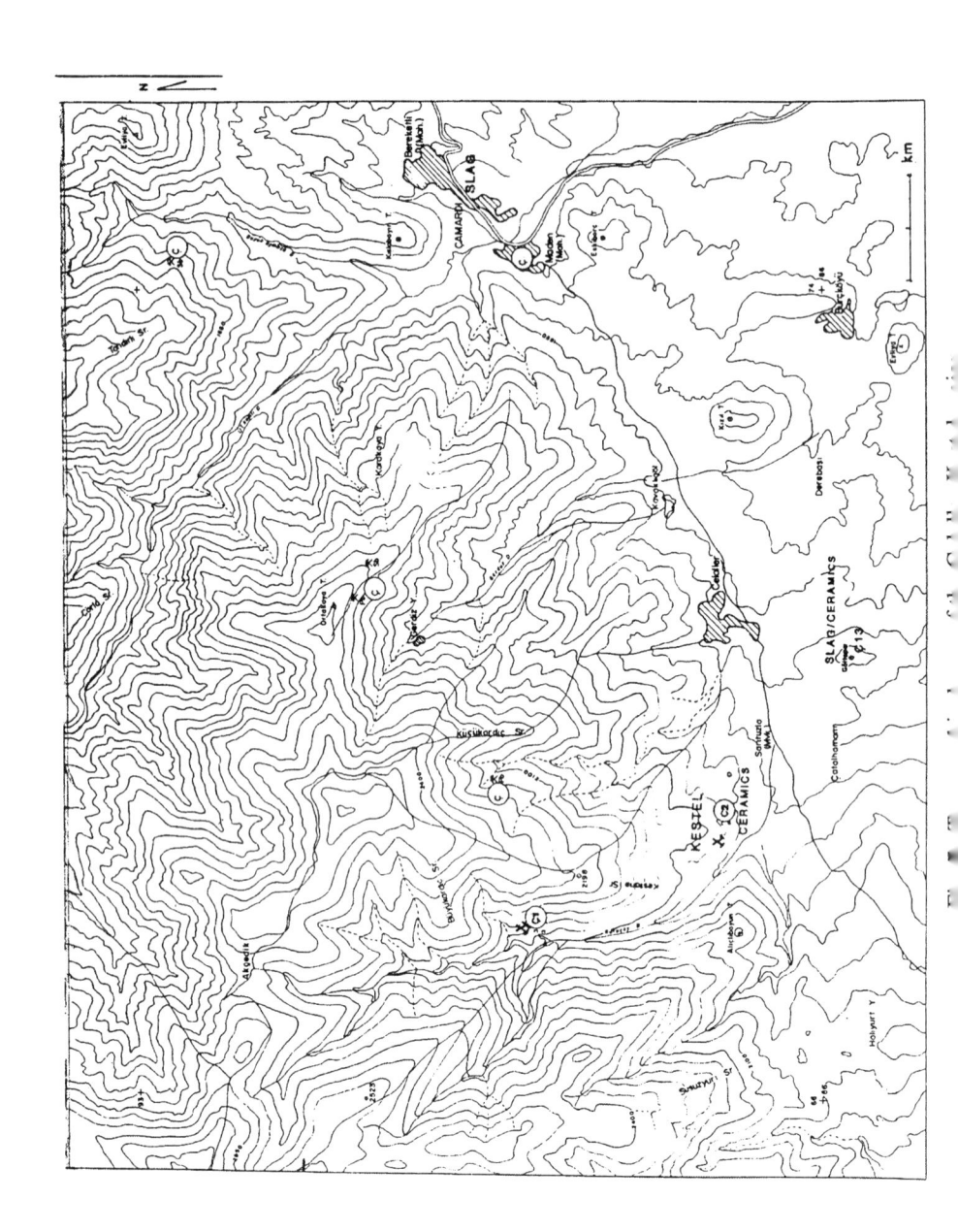

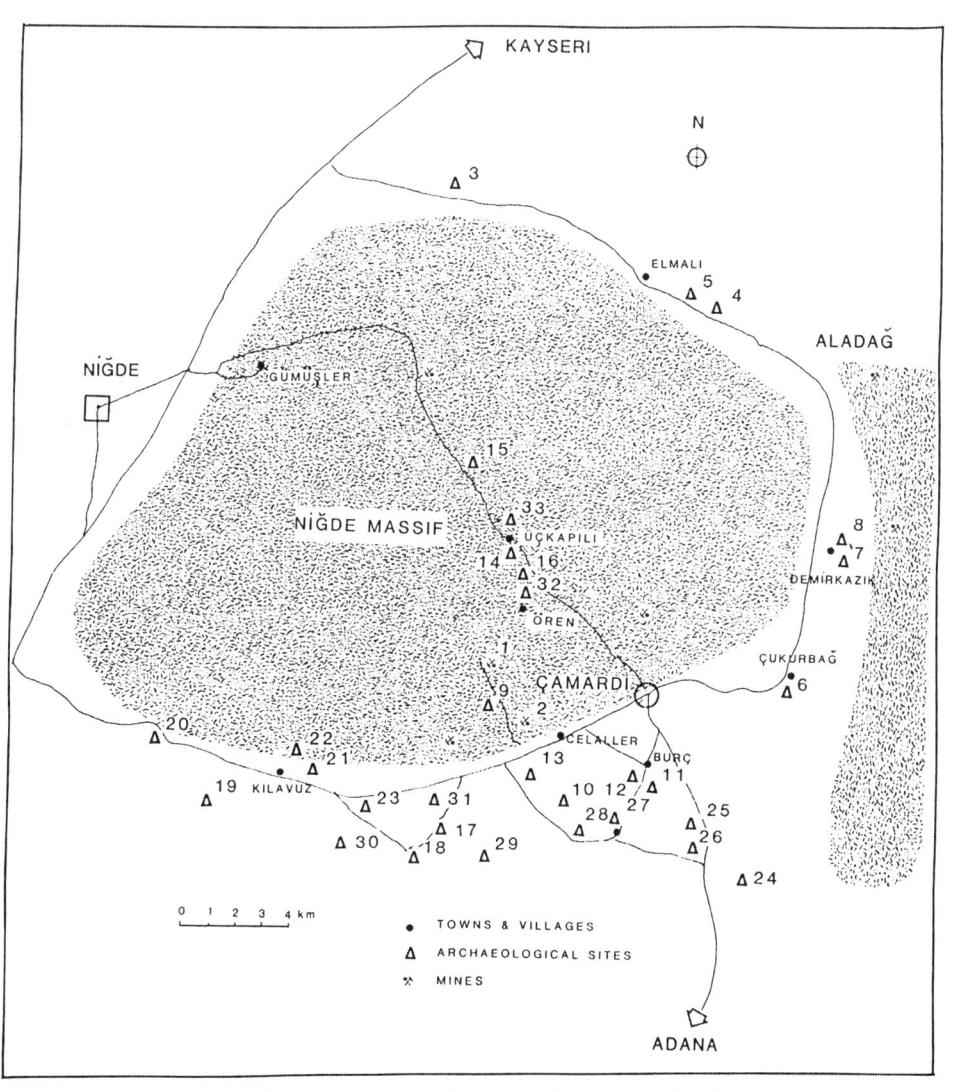

Fig. 8: Distribution of sites in the Niğde Massif

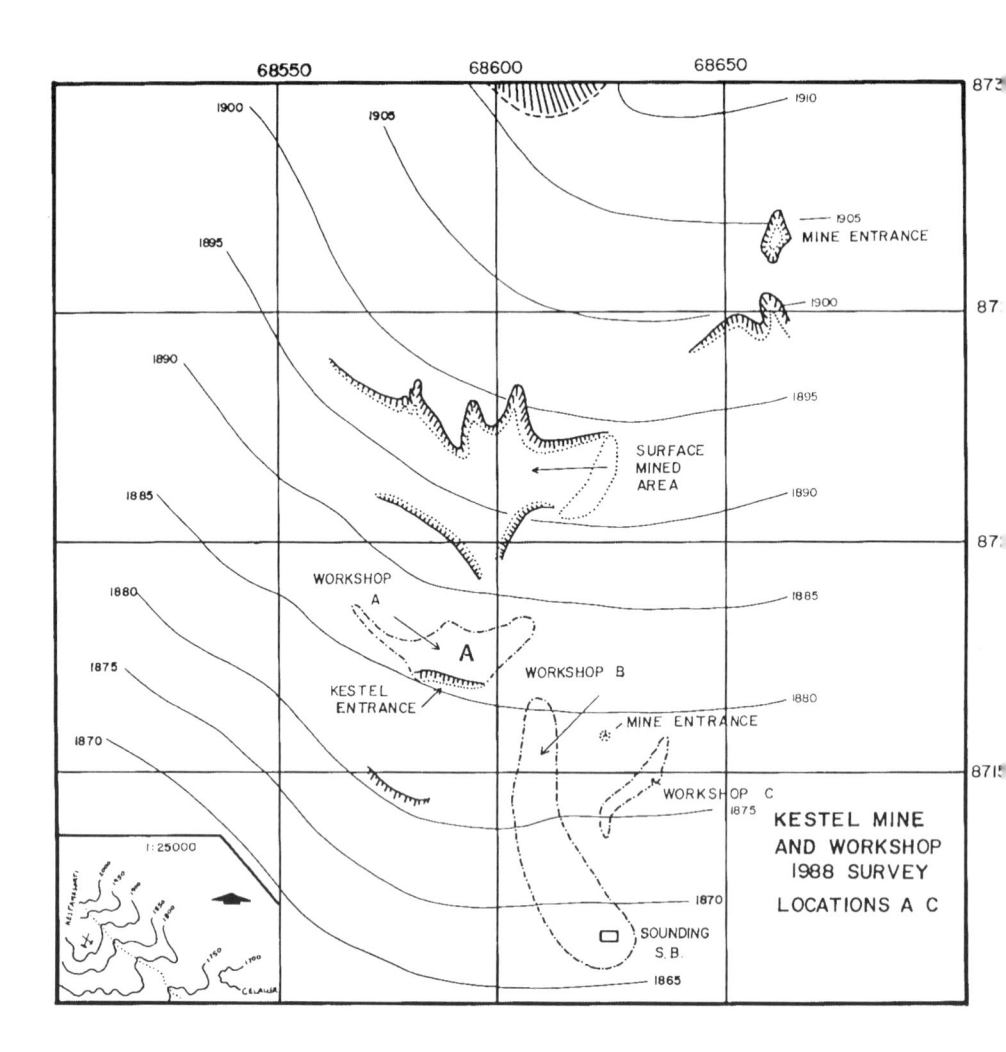

Fig. 9: Kestel mine slope (Sarıtuzla) workshop and mine entrances. 1988 Survey

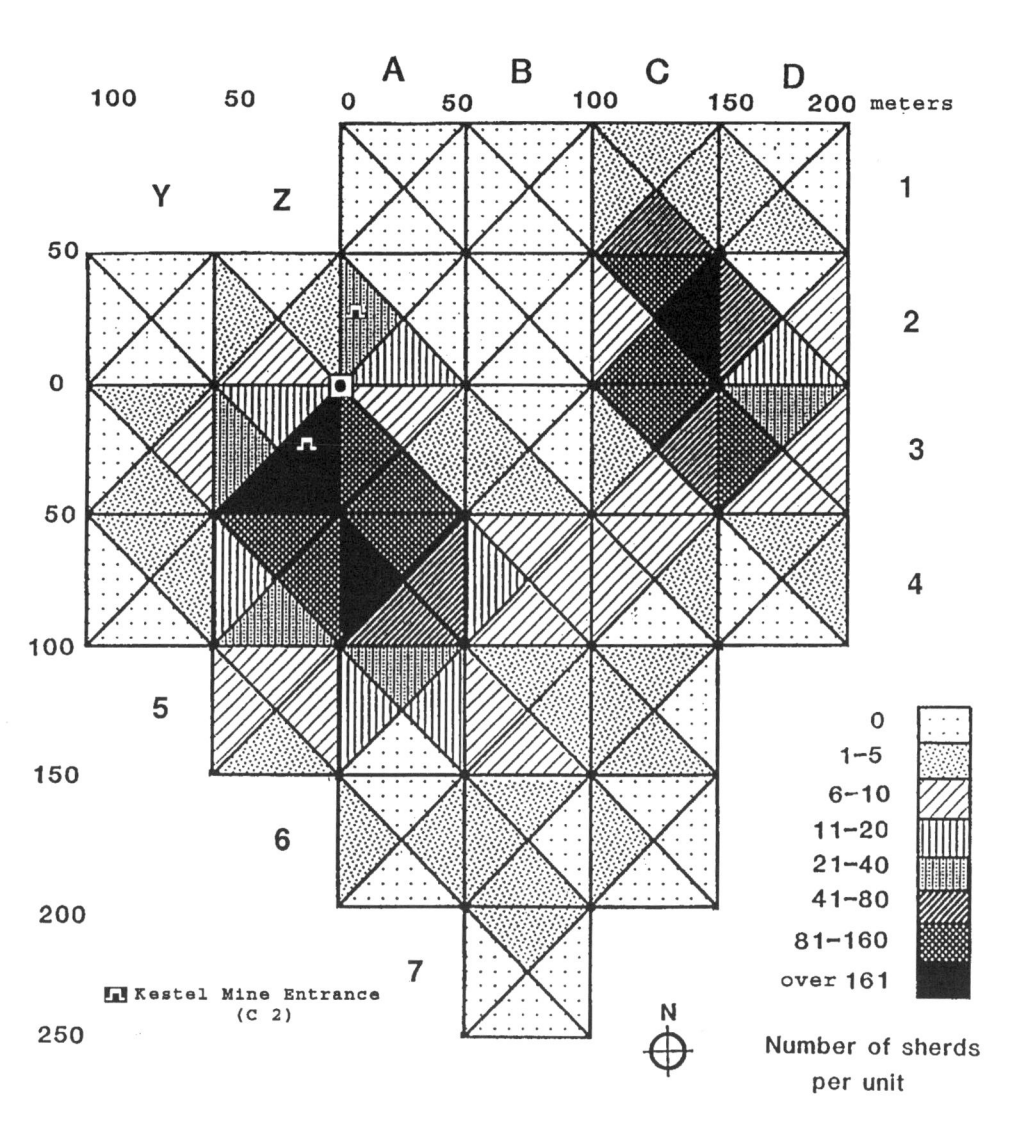

Fig. 10: Density map of ceramics from Kestel mine slope survey

Fig. 11: Large ore dressing installation at the roof of Kestel mine

FIGURES

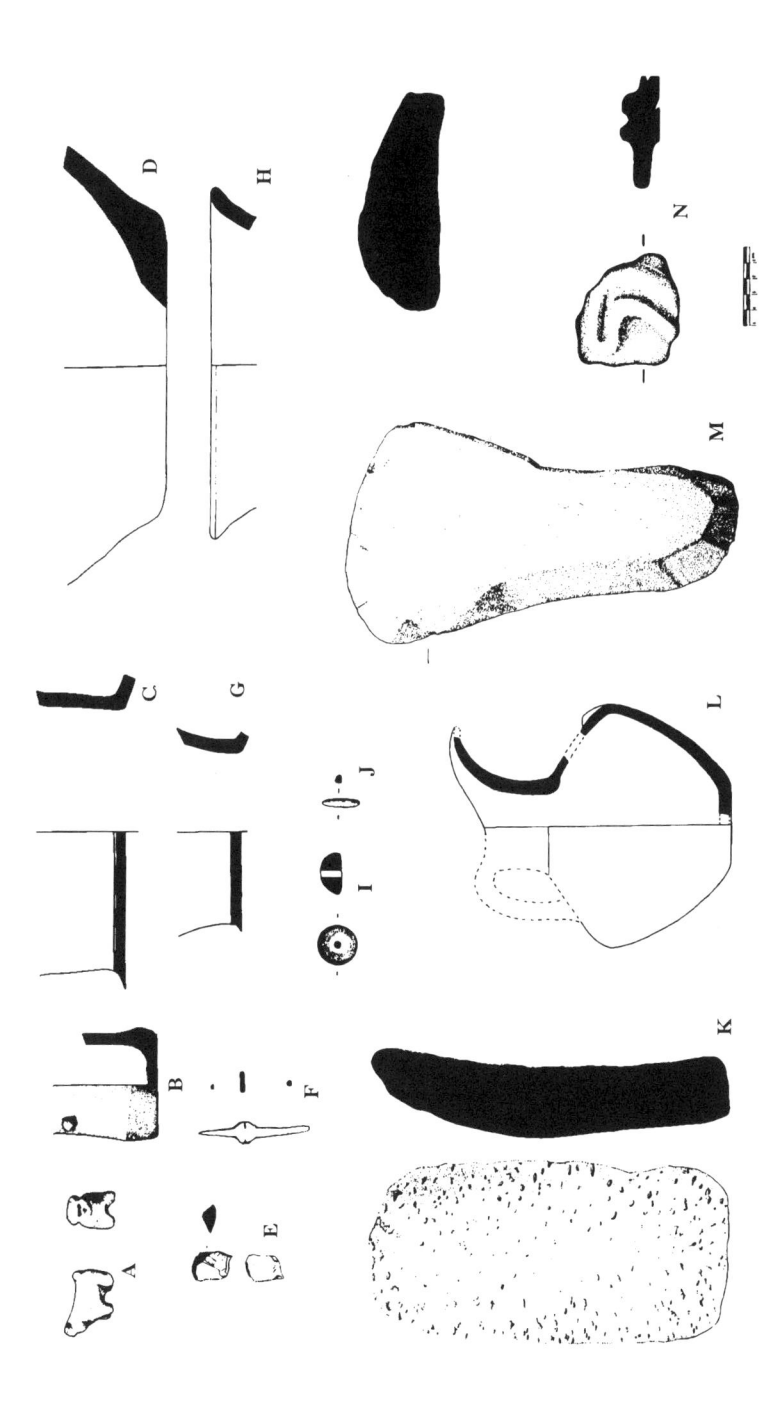

Fig. 12: Artifacts from Göltepe and Kestel Mine

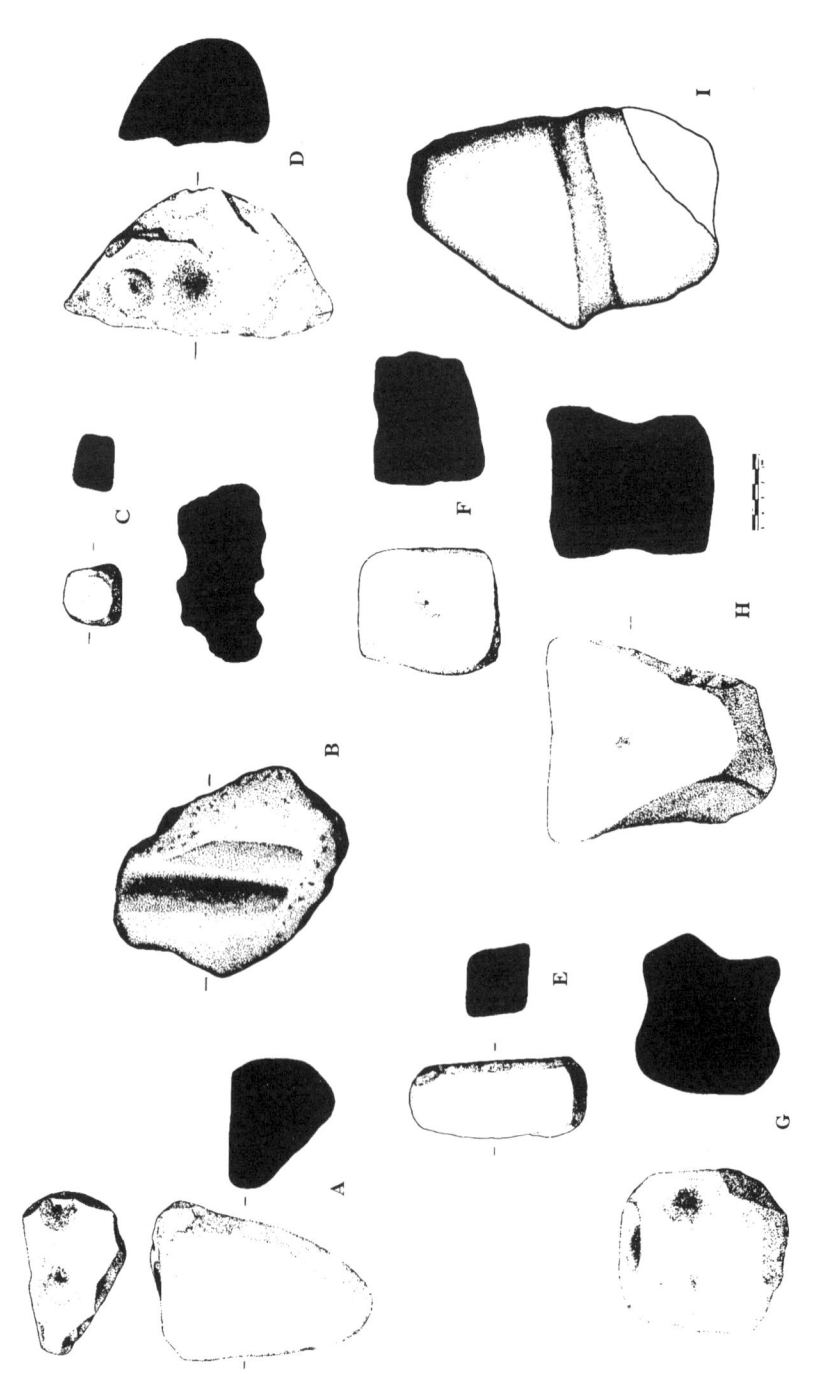

Fig. 13: Artifacts from Kestel Mine; scale applies to all artifacts except D, which is approximately 50 cm. in length

FIGURES 197

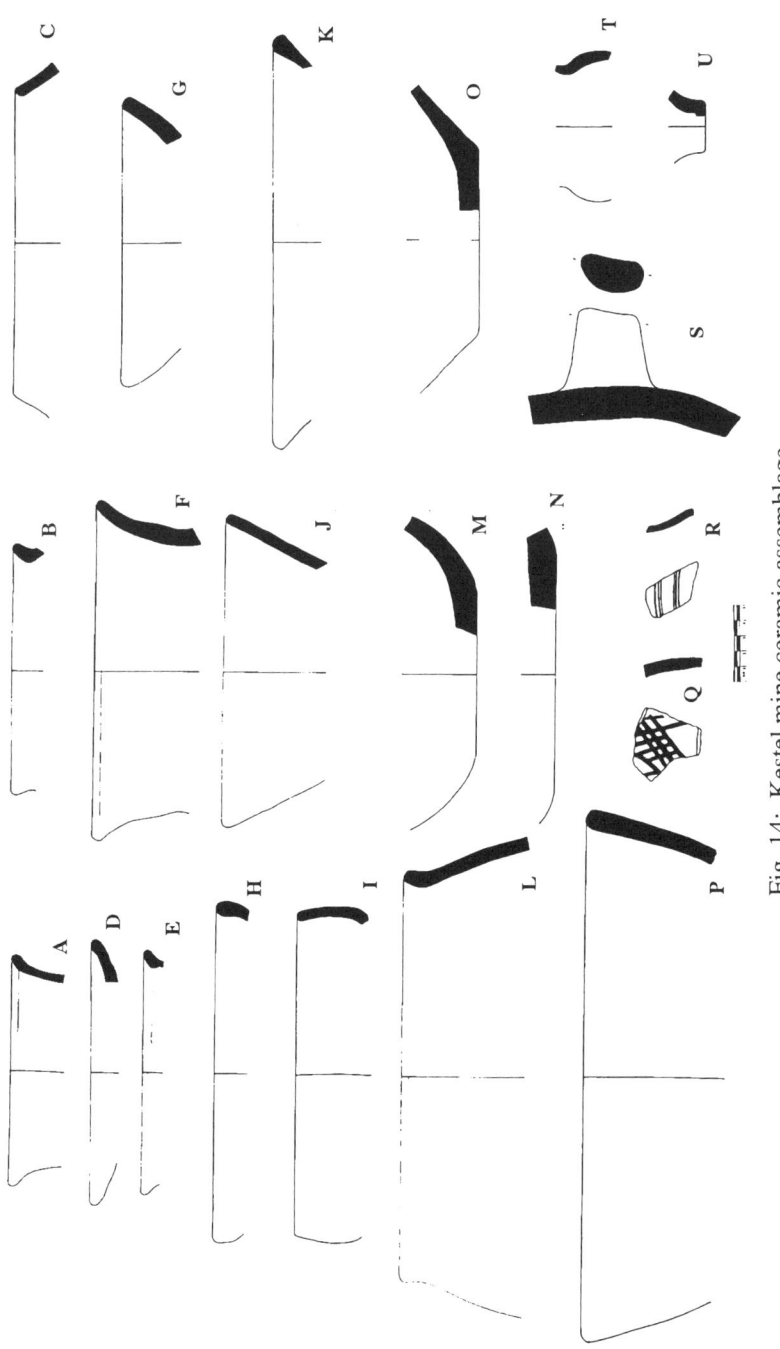

Fig. 14: Kestel mine ceramic assemblage

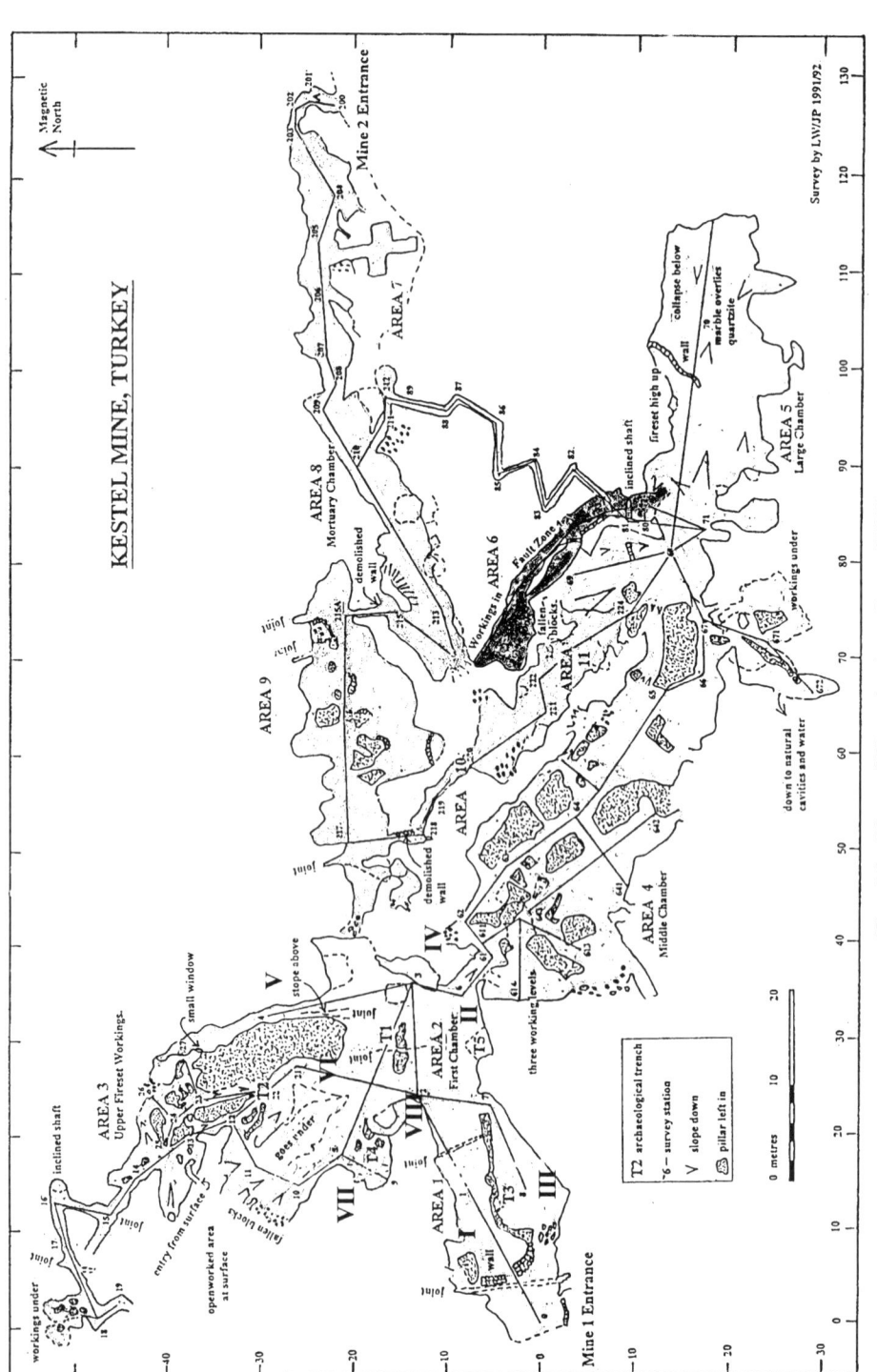

Fig. 15: Plan of Kestel mine. Lynn Willies

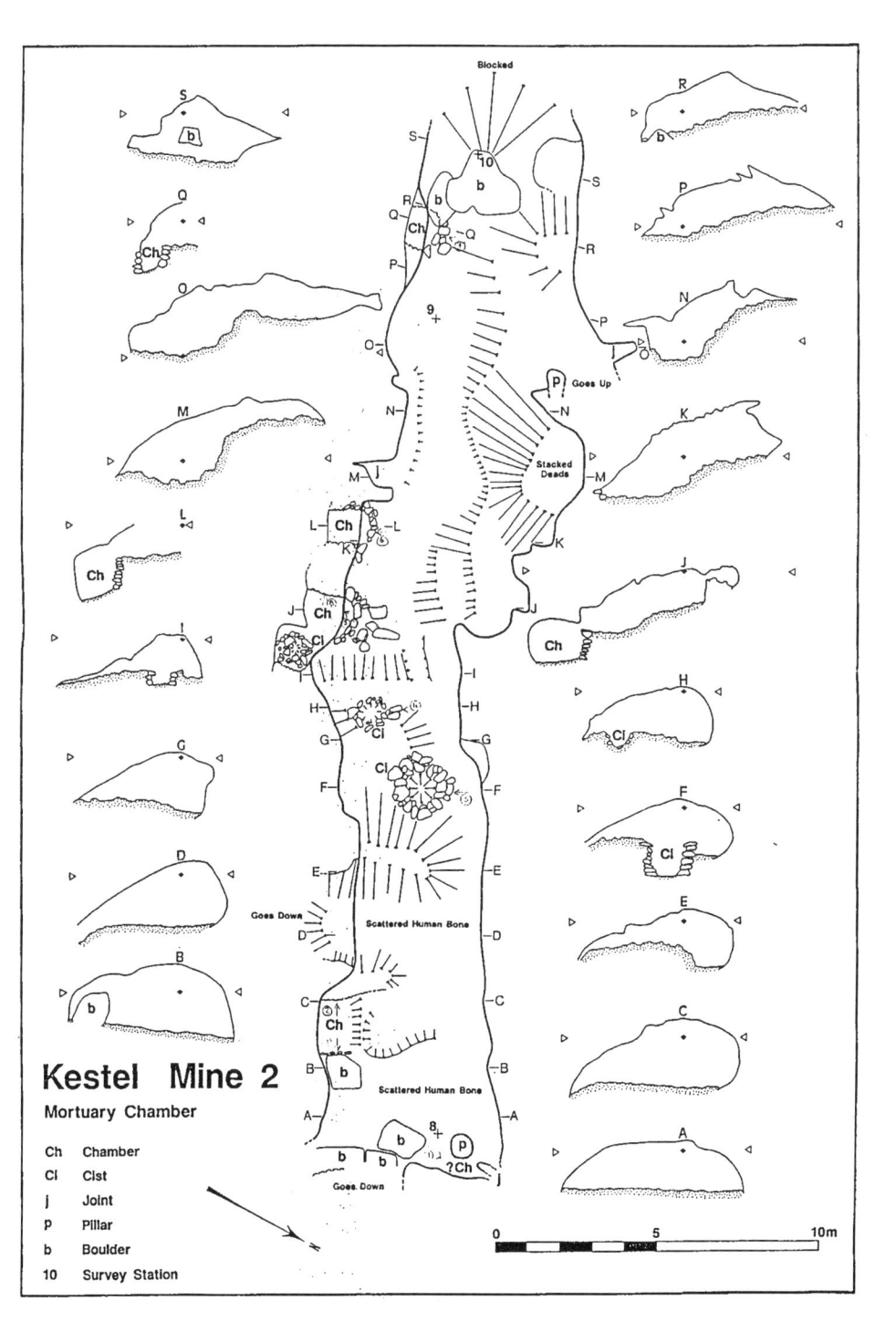

Fig. 16: Plan of Mortuary chamber, Kestel mine

200 FIGURES

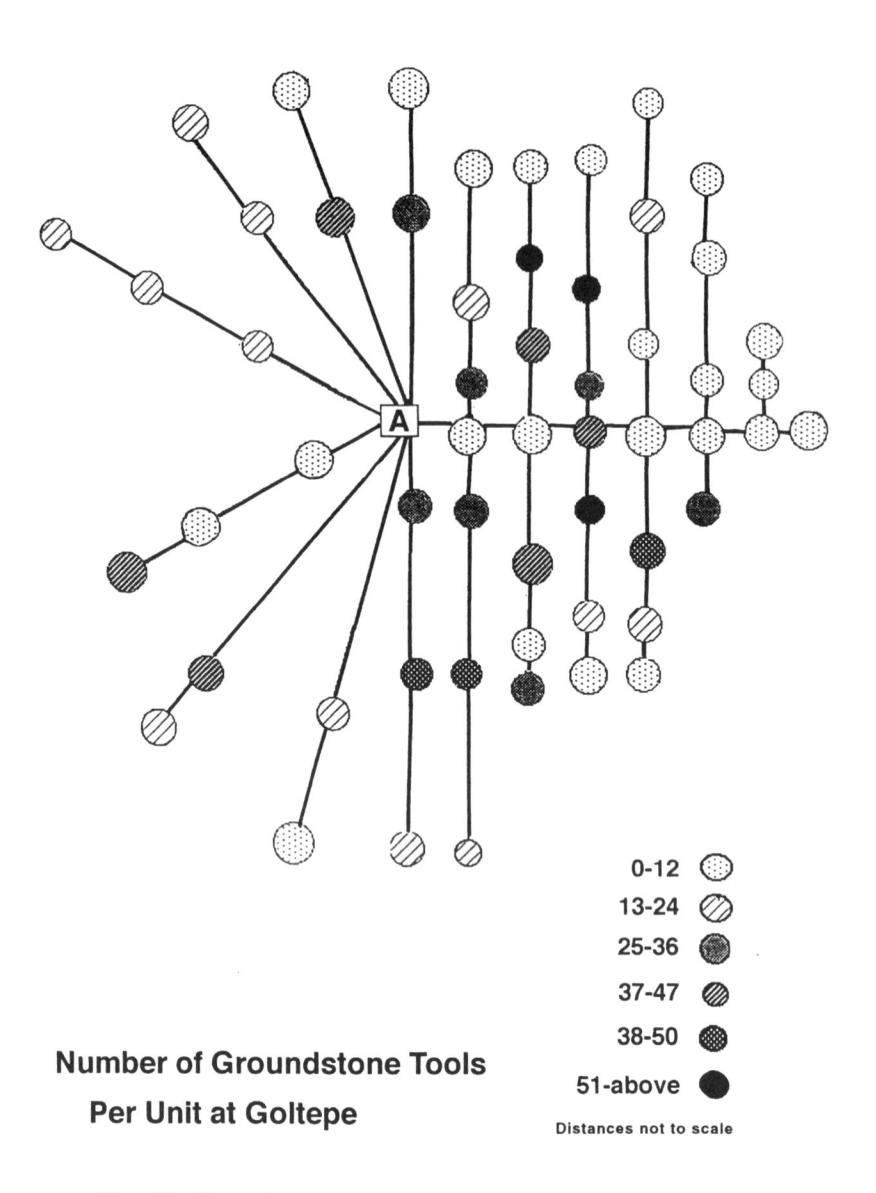

Number of Groundstone Tools Per Unit at Goltepe

0-12
13-24
25-36
37-47
38-50
51-above

Distances not to scale

Fig. 17: Groundstone tool distribution map. Göltepe survey

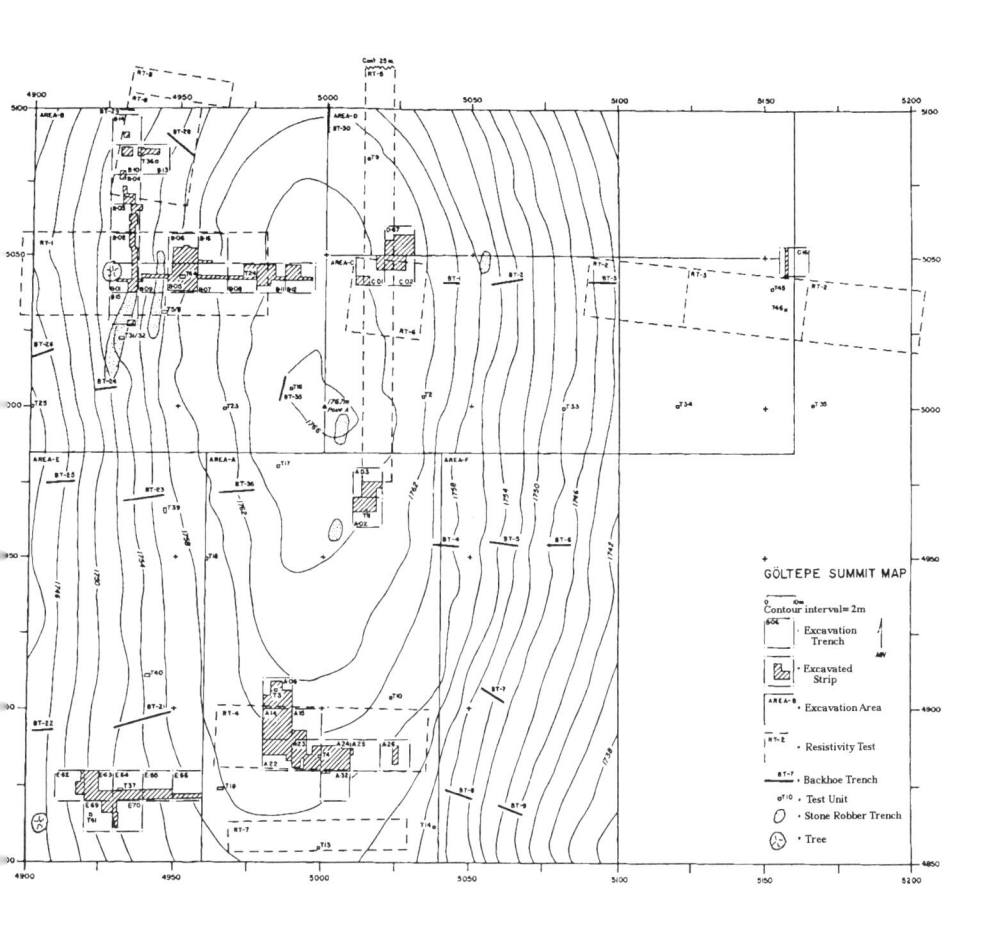

Fig. 18: Summit map of Göltepe

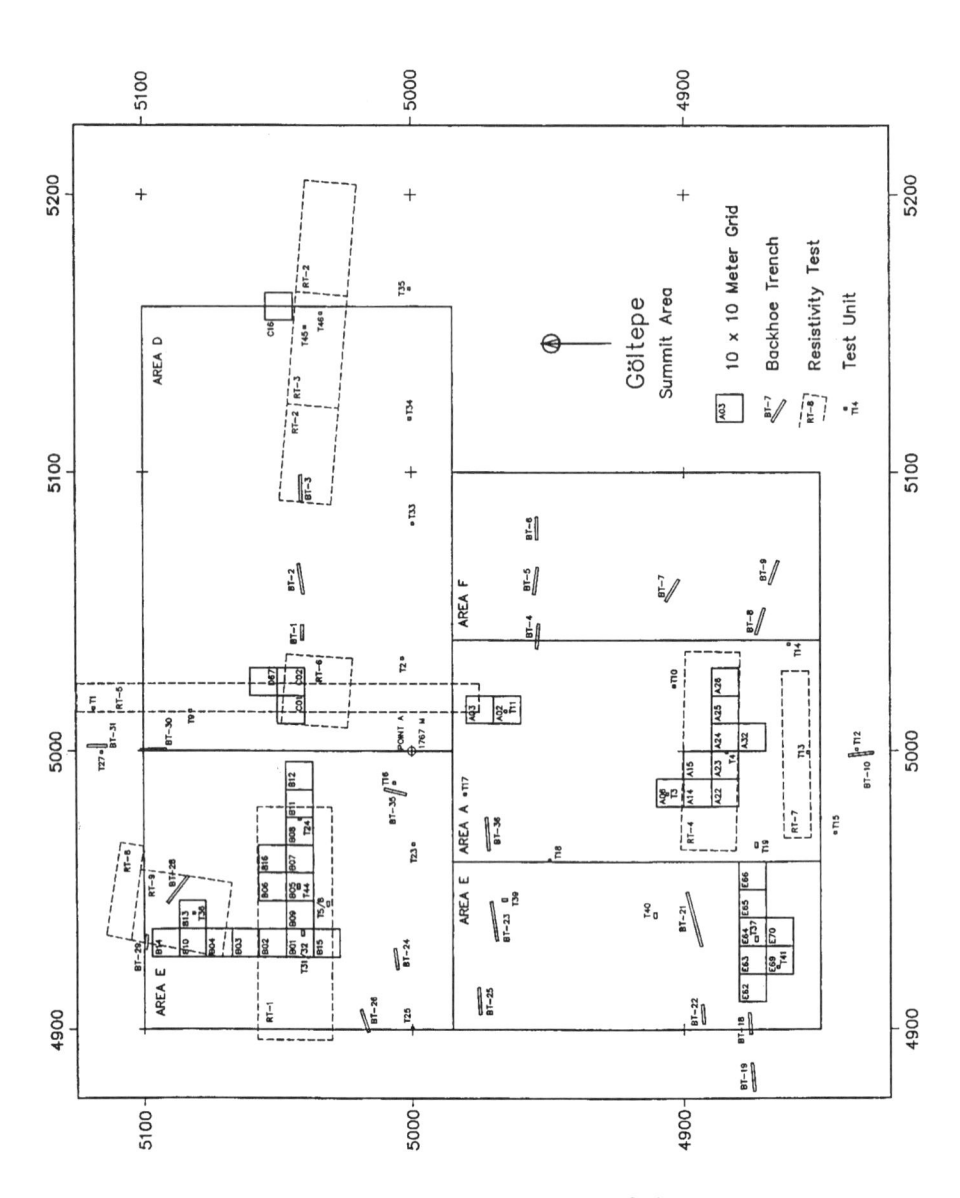

Fig. 19: Excavation trench map. Göltepe

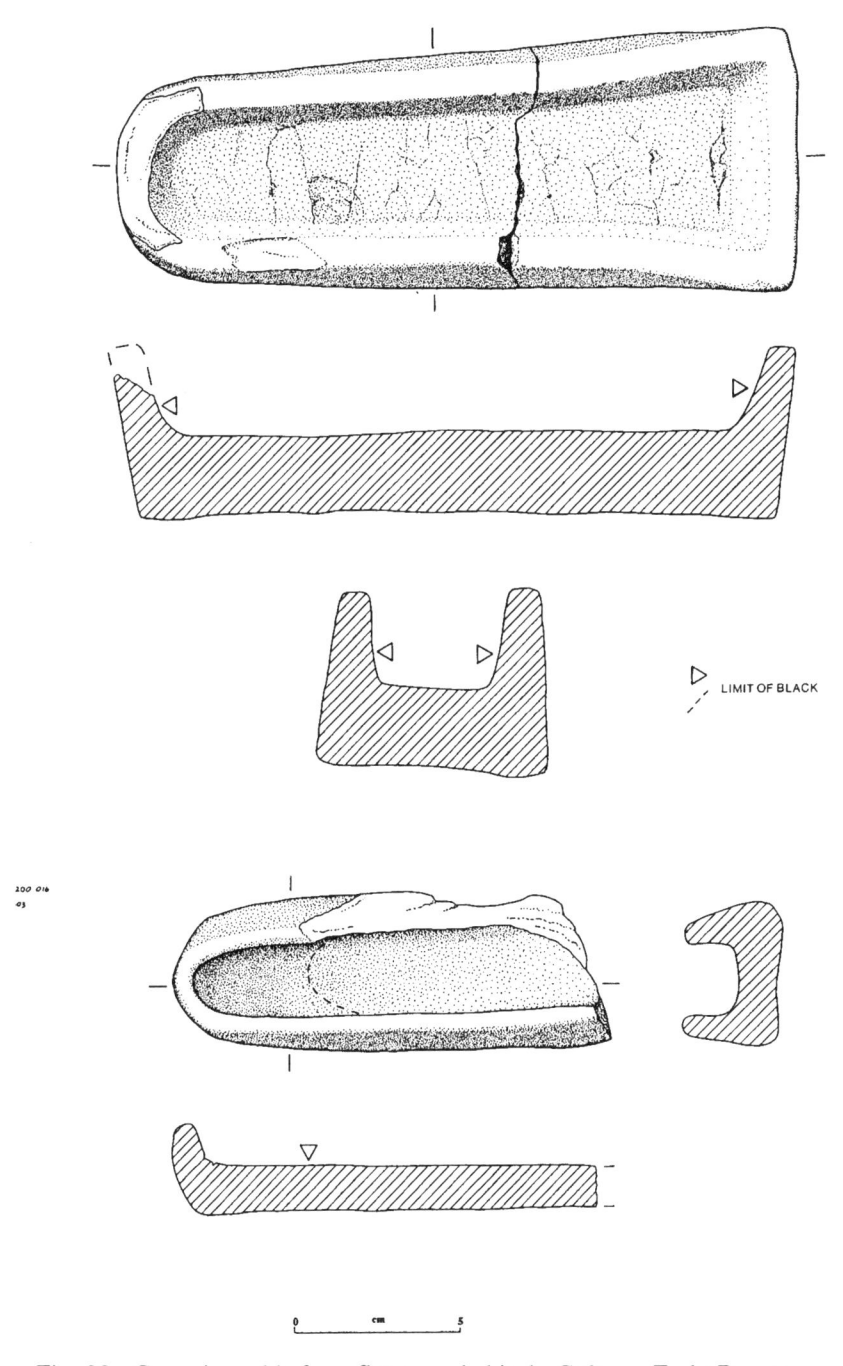

Fig. 20: Ceramic molds for a flat ax and chisel. Göltepe, Early Bronze Age

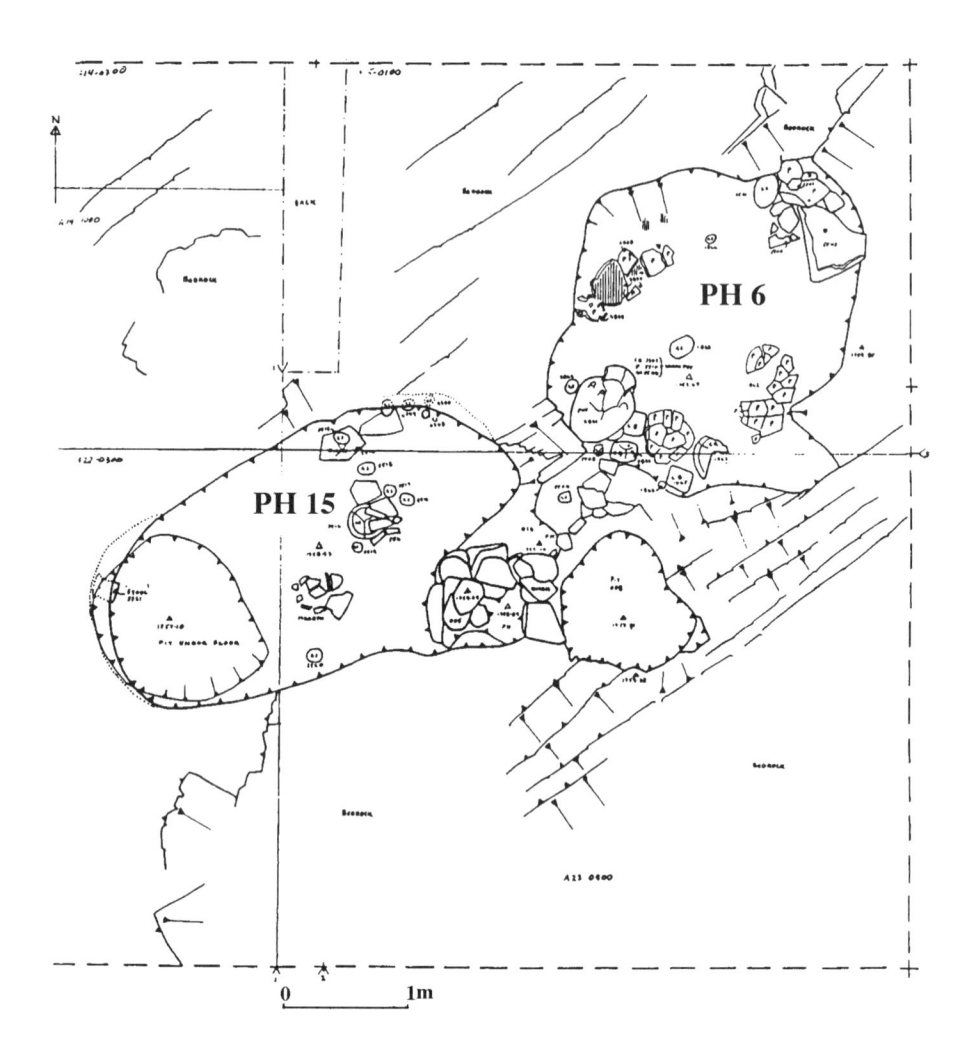

Fig. 21: Pithouse structures 6 and 15. Göltepe, Early Bronze Age

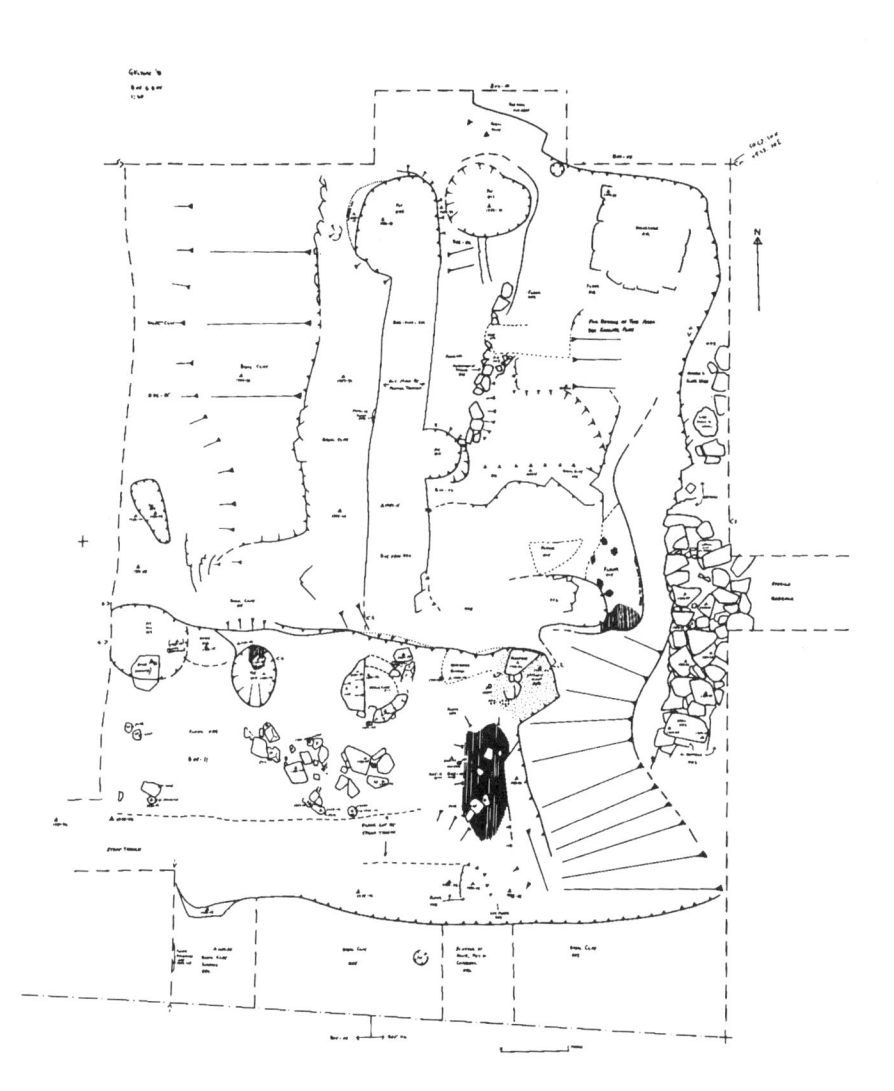

Fig. 22: Structures B05 and B06 in Area B. Göltepe, Early Bronze Age

Fig. 23: Silver necklace. Area B. Gőltepe, Early Bronze Age

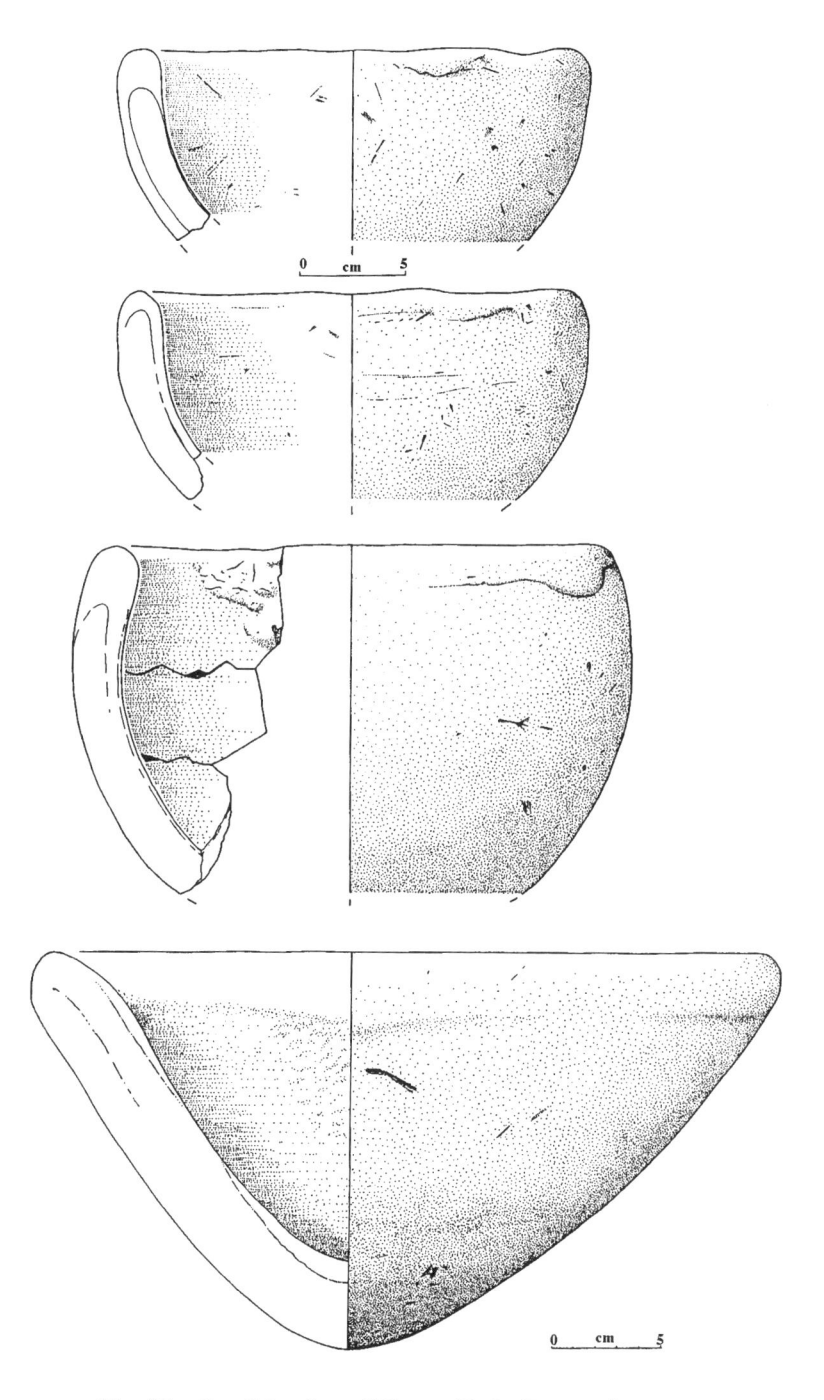

Fig. 24: Crucibles from Göltepe, Early Bronze Age

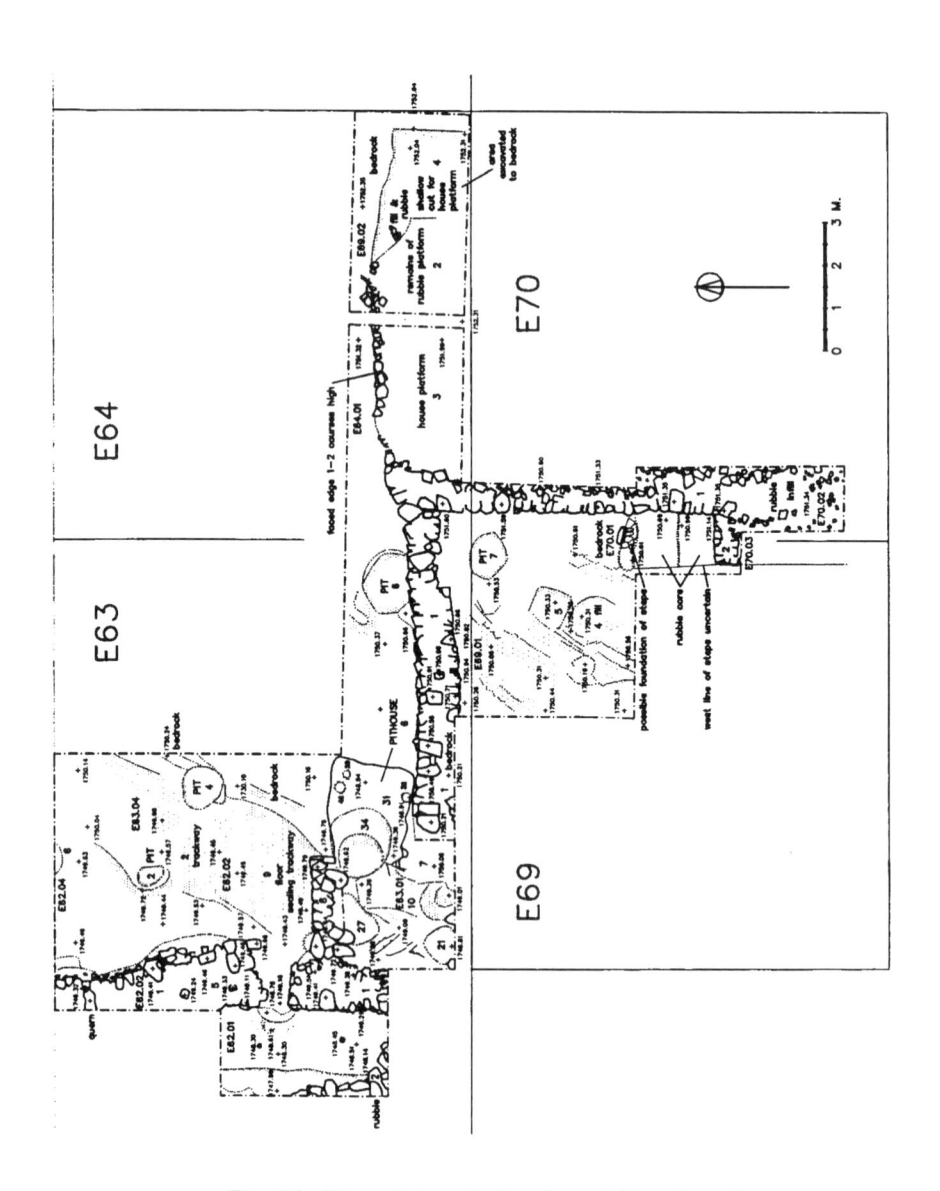

Fig. 25: Plan of Area E showing midden deposits

Fig. 26

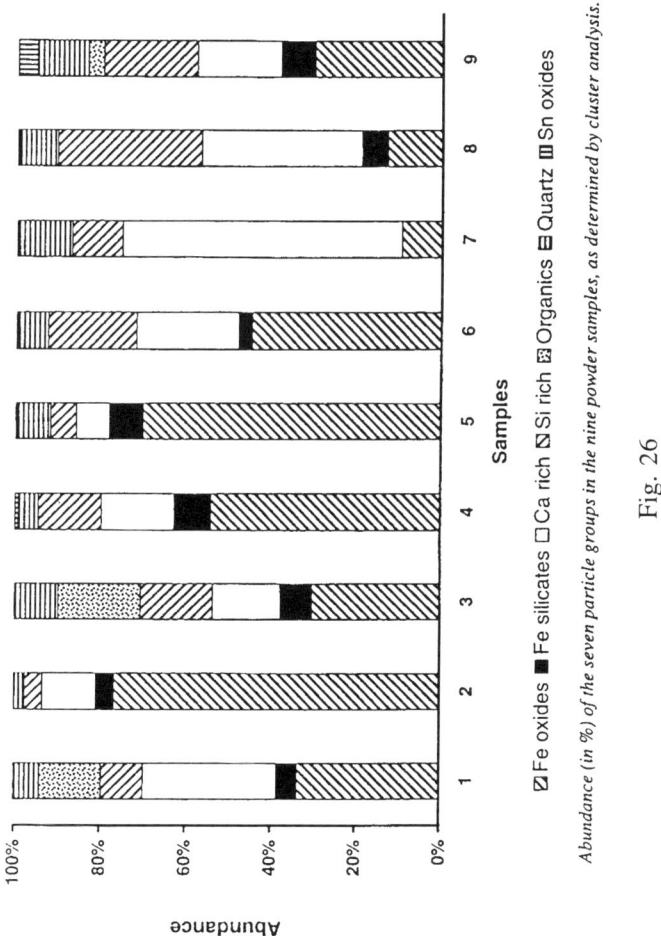

Abundance (in %) of the seven particle groups in the nine powder samples, as determined by cluster analysis.

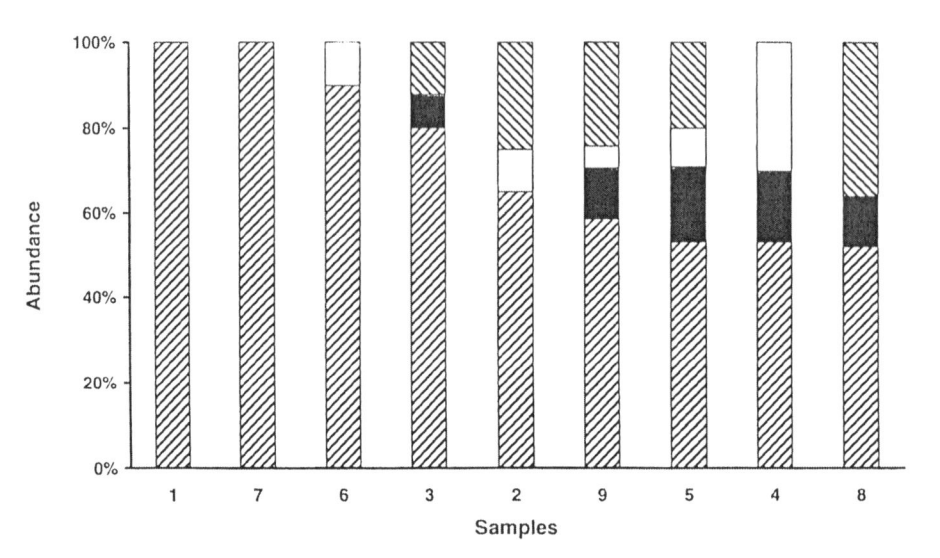

☒ tin oxides ■ tin silicates ☐ Fe-Sn rich ◨ Sn-Fe rich

Abundance (in %) of the four tin-containing particle groups in the nine powder samples, as determined by cluster analysis.

Fig. 27

PLATES

Plate 1: Areal view of Cilicia, Bolkardağ and Çamardı. December 16, 1972, M.T.A.

Plate 2: Cassiterite grains panned out of the Kuruçay Stream, Necip Pehlivan, M.T.A. 1987.

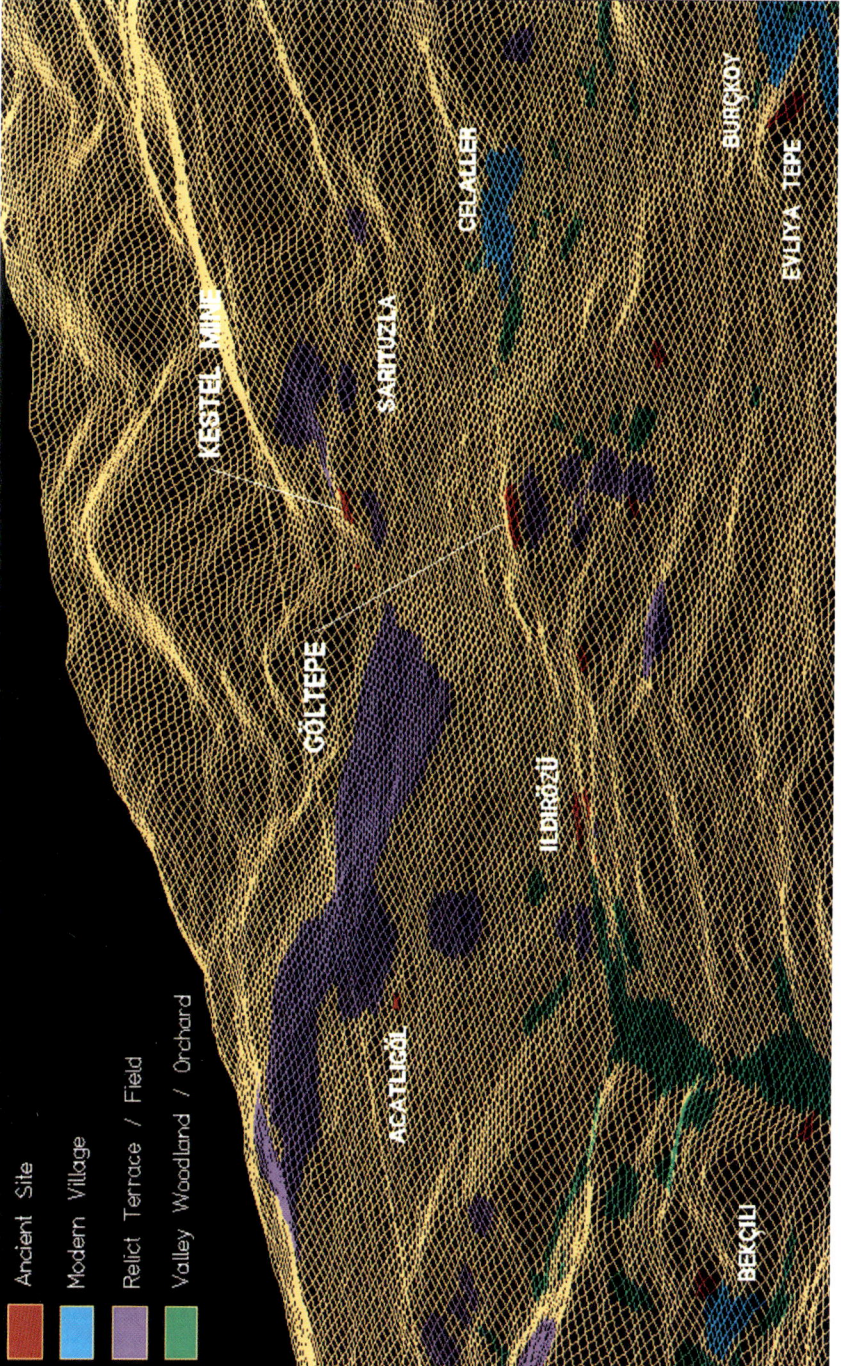

Plate 3: Computer model of Çamardı, Celaller, Kestel Mine and Göltepe region. Sanders, Oriental Institute.

Plate 4: Crucible fragments from Kestel slope area.

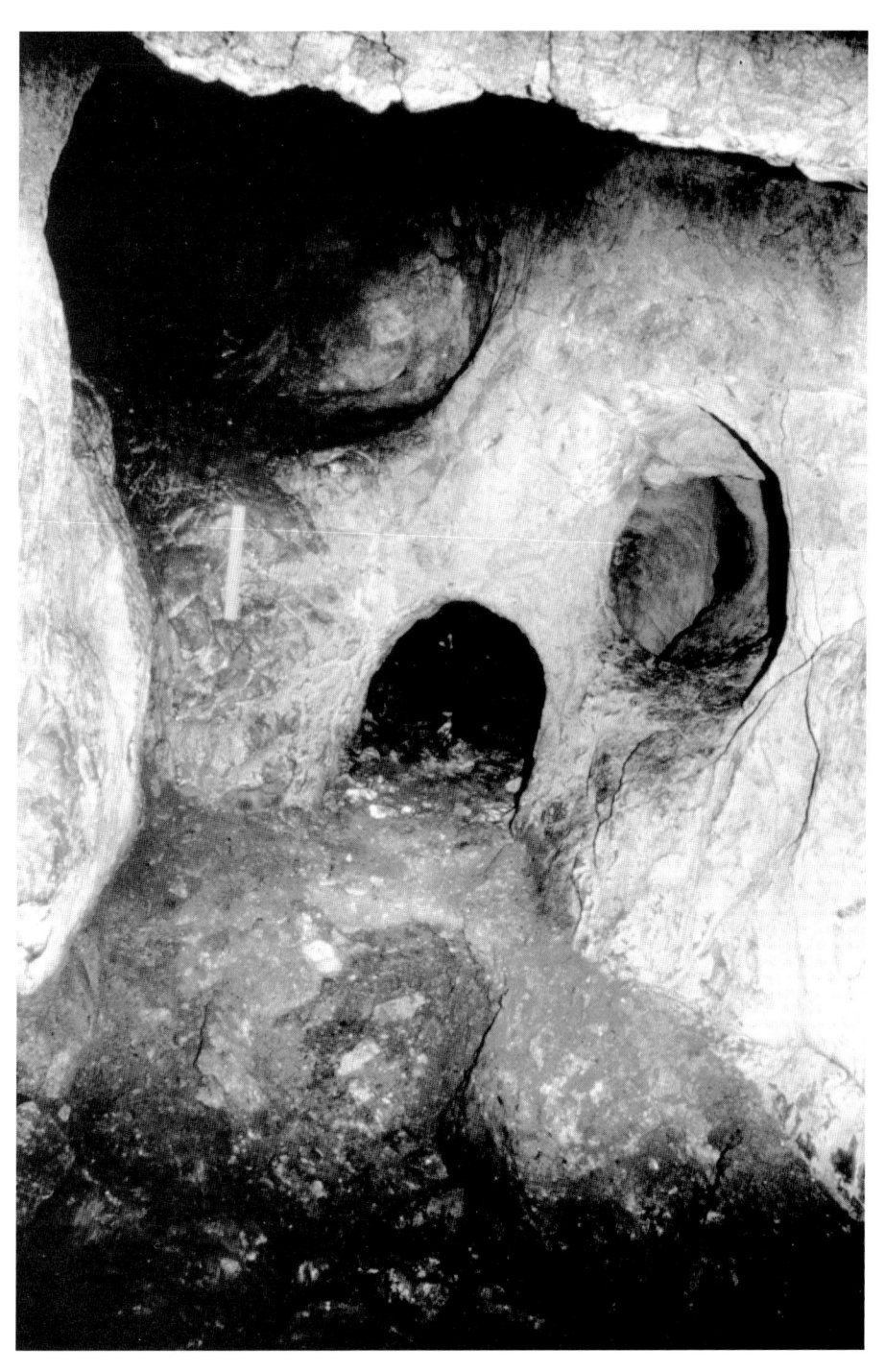

Plate 5a: Chamber VI Kestel Mine.

Plate 5b: Cassiterite grain from Sounding S2, Chamber VI, Kestel Mine.

Plate 6b: Chamber VI, Sounding 2, dark burnished ware.

Plate 6a: Chamber VI, Sounding 2, straw tempered ware.

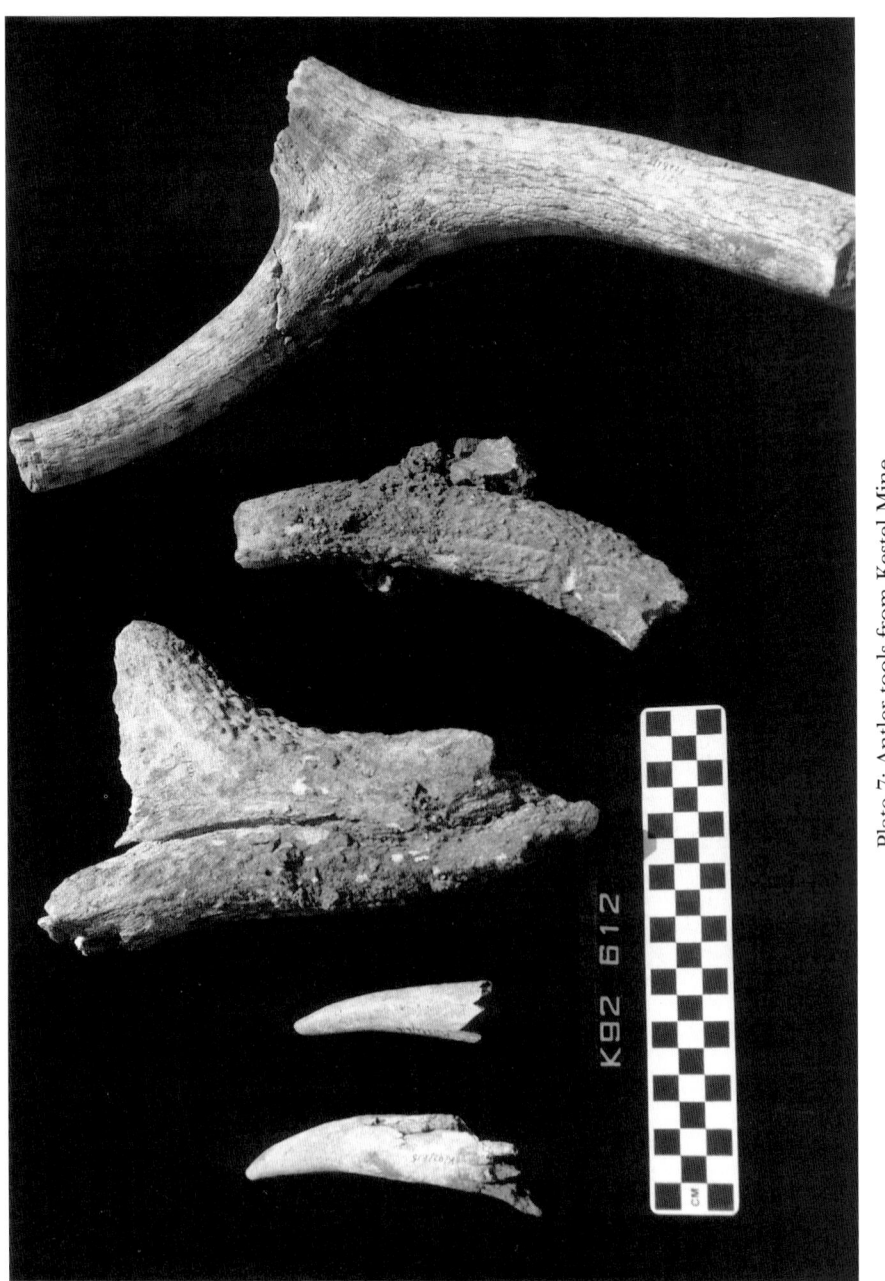

Plate 7: Antler tools from Kestel Mine.

Plate 8: Decorated ceramic panel Pithouse 6, Göltepe, Early Bronze Age.

Plate 9: Large storage vessel containing ground ore material, Pithouse 6, Göltepe, Early Bronze Age.

Plate 10: Large crucible with stone covers on floor of Pithouse 15, Göltepe, Early Bronze Age.

Plate 11: Geometrically decorated ceramic panel over hearth. Structure B05, Göltepe, Early Bronze Age.

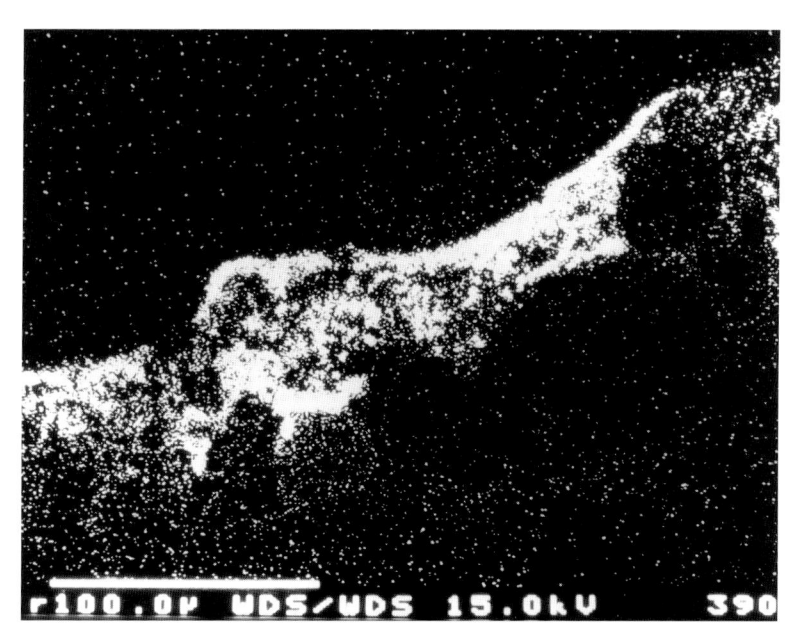

Plate 12: Tin x-ray map of Crucible MRN 537 cross-section. SIMS. Mieke Adriaens.

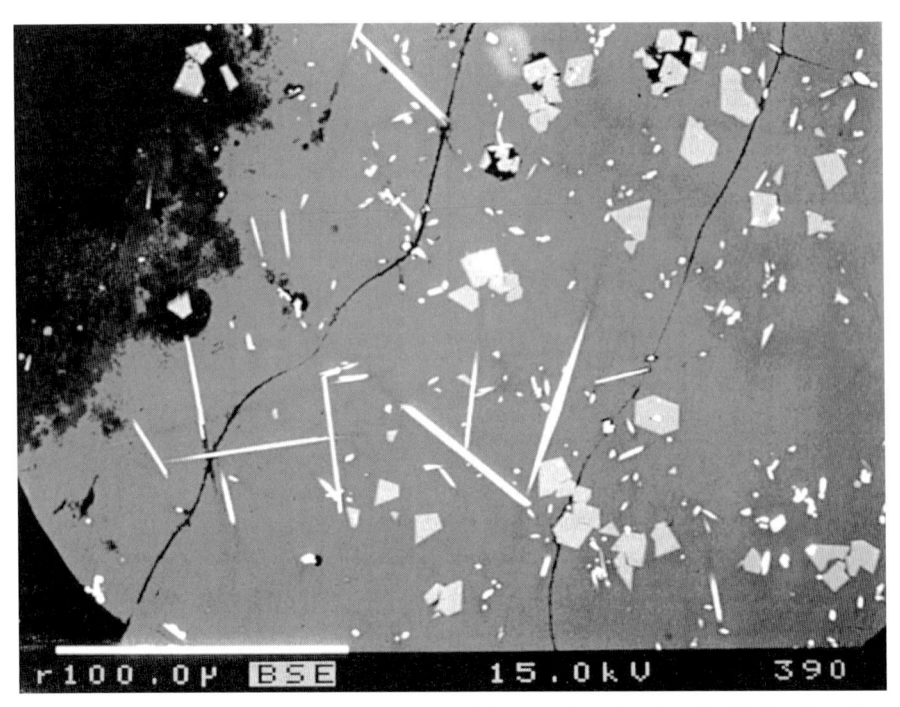

Plate 13: Microprobe image of glassy crucible accretion. Ian Steele, University of Chicago.

analyses of Sn-bearing samples. Ian M. Steele

	----- Fe-Sn crystals ------			---------- Matrix --------			Sn crystal
SnO2	18.8	18.2	17.1	15.2	13.4	13.0	97.5
CaO	0.09	0.06	0.06	16.6	16.8	17.2	0.36
MgO	2.30	2.33	2.07	1.81	1.90	1.82	0.28
Al2O3	2.69	2.50	2.31	5.83	6.11	5.86	0.61
SiO2	0.0	0.0	0.0	32.7	32.6	33.9	1.01
Na2O	0.0	0.0	0.0	1.05	1.05	0.97	0.10
K2O	0.02	0.01	0.01	2.68	2.57	2.70	0.25
FeO	69.9	71.0	71.7	21.4	21.7	21.2	2.02
Total	93.8	94.1	93.3	97.3	96.1	96.7	102.1

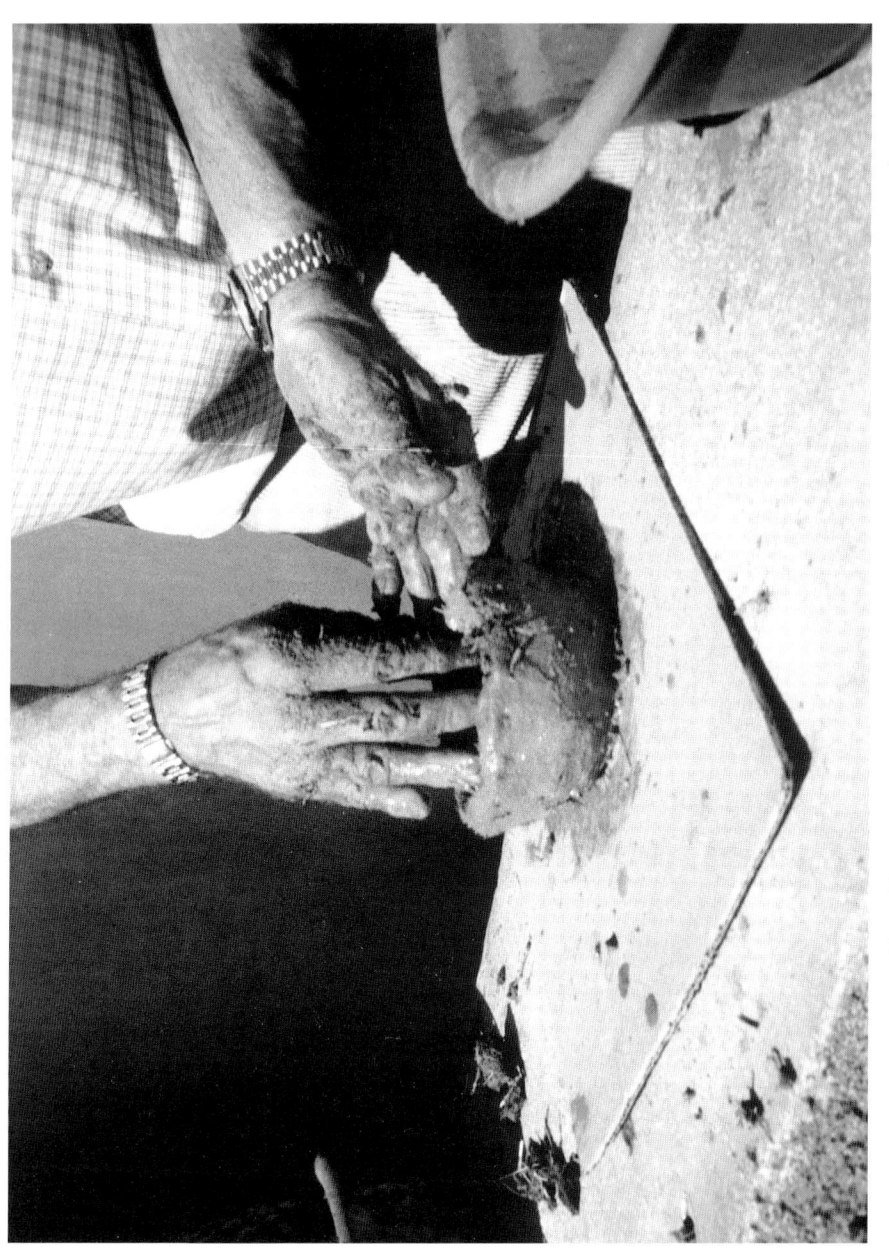

Plate 14: Replication of crucibles with Celaller clay. Bryan Earl.

Plate 15: Ore beneficiation, vanning with a shovel, Oriental Institute courtyard experimental smelt. Bryan Earl.

Plate 16a: Tin metal prill from experimental smelt, using Göltepe ore materials. Experimental smelt.

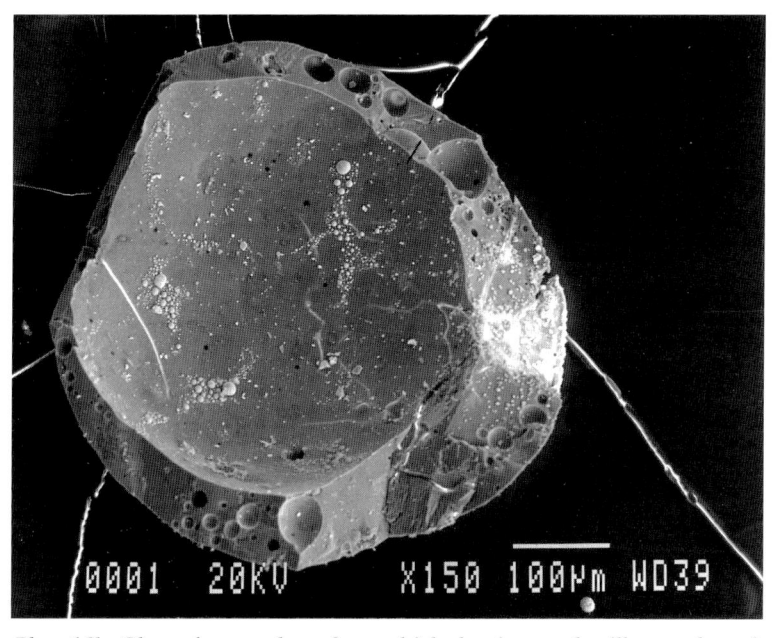

Plate 16b: Glassy slag envelope from which the tin metal prill was released upon grinding. Experimental smelt.

Plate 17: Experimental Smelt, Celaller Village using three blowpipes.

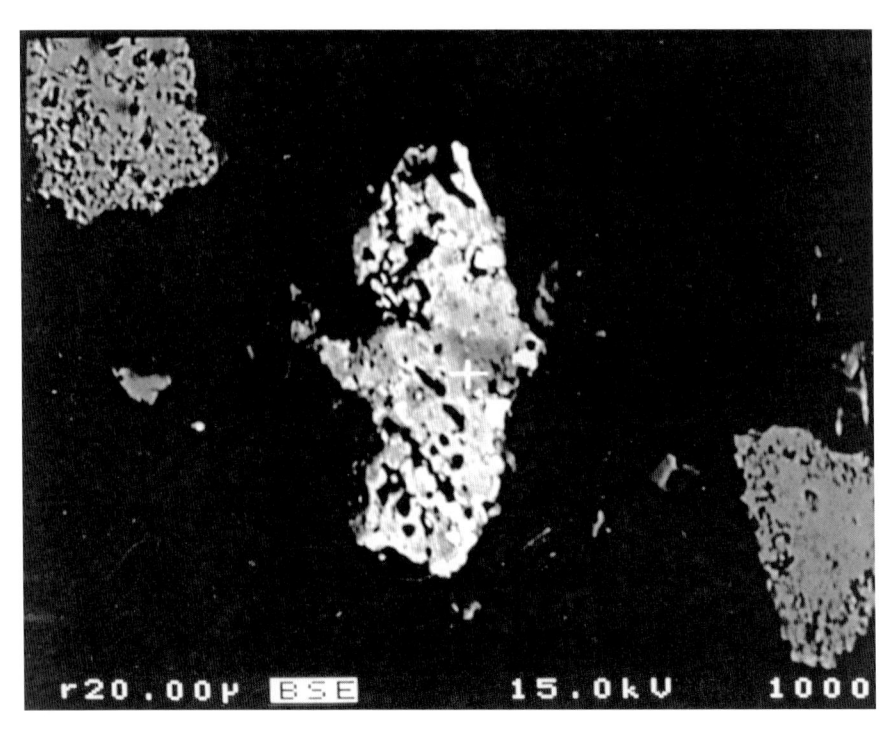

Plate 18: XPS metallic tin from powdery ore material, Göltepe, Early Bronze Age.
Mieke Adriaens.